INVESTIGATING THE FIREGROUND

CALVIN C. PHILLIPPS
Chief, City of Oshkosh Fire Department
Oshkosh, Wisconsin

DAVID A. McFADDEN
Instructor Coordinator, Fire Protection Program
Fox Valley Technical Institute
Appleton, Wisconsin

ROBERT J. BRADY CO.
A Prentice-Hall Publishing and Communications Company
Bowie, Maryland 20715

Executive Editors: George Post
Jim Yvorra
Production Editor: Janis K. Oppelt
Art Director: Don Sellers
Text Designer: Michael J. Rogers
Illustrator: George Post
Cover Photograph: Rick Brady
Photographs: All photographs supplied by the authors unless otherwise indicated
Typesetting: Prestige Editorial and Graphics Services, Inc., Washington, D.C.
Typeface: Century Schoolbook
Printed By: Fairfield Graphics, Fairfield, Pennsylvania

Investigating the Fireground

Library of Congress Cataloging in Publication Data

Phillipps, Calvin C., 1926-
Investigating the fireground.

Includes index.
1. Fire investigation. I. McFadden, David A., 1944- . II. Title.
TH9180.P48 363.3′764 82-1290
ISBN 0-89303-074-0 AACR2

Prentice-Hall International, Inc., London
Prentice-Hall of Australia, Pty., Ltd., Sydney
Prentice-Hall of India Private Limited, New Delhi
Prentice-Hall of Japan, Inc., Tokyo
Prentice-Hall of Southeast Asia Pte. Ltd., Singapore
Whitehall Books, Limited, Petone, New Zealand

Printed in the United States of America

82 83 84 85 86 87 88 89 90 91 92 10 9 8 7 6 5 4 3 2 1

CONTENTS

FOREWORD

The systematic search for fire cause is an important, yet often neglected, function of the fire service. Careful fire investigation can lead to the identification of preventable, accidental causes as well as the identification of arson.

Effective fire investigation requires not only a departmental commitment from the Chief to the first-in firefighter. There must also be adequate training, especially at the investigator level. For a fortunate few, this will be formal training, but for most it will be a combination of on-the-job trial and error plus independent study.

This book can assist those who must rely upon independent study. It is written by firefighters for firefighters in no nonsense terms. The technical data on fire behavior, cause, and origin is presented clearly. The investigative processes describe a logical sequence which, if followed, lead the investigator through a comprehensive investigation. The case histories reinforce many important points.

The authors' intent was to provide sufficient information to conduct a basic fire investigation. However, the veteran investigator will also find this book a useful reference to turn to time and again in difficult or unusual cases.

As in most endeavors, simply reading about fire investigation will not make the student proficient. No amount of reading prepares the investigator for the inadequacy we experience when confronted with a devastating fire and told to "find the cause."

Coach Vince Lombardi once observed, "Practice does not make perfect; perfect practice makes perfect." Hopefully, this book will assist those who strive for perfect practice.

Philip C. Culp
Director, Arson Bureau
Wisconsin Division of Criminal Investigation
Madison, Wisconsin

PREFACE

Arson has been identified as one of the fastest growing crimes in America. Proper fire investigation is imperative to uncover this crime and in many cases, if the potential arsonist is aware that a thorough investigation *will* be done, arson can be deterred.

Many aspects of fire investigation are covered in this text including how to determine the point of origin, heat source, reason, and category. The step-by-step process of conducting an investigation is reinforced with case histories which provide examples of a variety of fire situations and illustrate how to use each step. Results of such an investigation can be used in training firefighters about newly uncovered hazards, in public education programs, and in fire prevention programs.

The science of fire investigation requires the cooperation of a wide range of professionals. This text has been written for firefighters, fire officers, fire investigators, instructors, insurance personnel, attorneys-at-law, law enforcement personnel, and military personnel.

The authors wish to acknowledge the support in this endeavor of the following: Captain Gary Kaufman, Training Officer, Oshkosh Fire Department; the City of Oshkosh Fire Department; and the Town of Menasha Fire Department.

We also wish to thank George Post for his patience and motivation in the writing of this text.

We are grateful to Ed Millman, technical writer, for his assistance in the writing and editing of this text.

Calvin Phillipps
David A. McFadden

DEDICATION

To our sincere friend, George D. Post, whose unselfish assistance and contributions have enhanced immeasurably the quality of training, education, and proficiency of the fire service.

1

FIREGROUND INVESTIGATION

Arson is the crime of deliberately setting fire to property for fraudulent or malicious purposes. As with any crime, the purpose, or motive, is important. For example, a homeowner may deliberately set fire to a pile of leaves in his backyard. His motive is to get rid of the leaves as simply and effectively as possible; no crime is involved. But when a person deliberately sets fire to his own property to collect on the insurance policy, or sets fire to someone else's property in a fit of anger, then the crime of arson has been committed.

The incidence of arson fires and fires suspected to be arson has been increasing over the past two or three decades—both in number and as a percentage of all fires. Until recently, anti-arson efforts consisted mainly in the investigation, by arson squads and/or fire marshals, of fires that were obviously deliberate. Some arsonists were detected and convicted, but many were not. These efforts alone did not prove to be a deterrent to the fraudulent or malicious setting of fires.

In the last several years, a number of municipalities have instituted wide-ranging programs to combat arson. The reason for such programs is obvious: Arson fires result in deaths and injuries to civilians and firefighters, property losses and the wasting of public funds, loss of jobs and income, and many other varieties of physical, mental, and economic hardship.

Each anti-arson program is tailored to fit the particular needs and situation of the community it serves, and may include some or most of the following activities:

1. Immediate investigation and documentation of every fire by fire department personnel, to determine the category of the fire and the possibility of arson

2. Formation of an arson investigation squad with police powers (if one does not already exist) to perform criminal investigations of fires suspected to be arson
3. Anti-arson legislation, to remove the profit from arson, give arson investigation the same legal basis as other types of criminal investigation, and provide increased penalties for arson
4. Greater cooperation between fire and police departments (and, in some cases, insurance companies) in deterring and investigating arson
5. Public awareness campaigns, to explain the problem to the general public and enlist their help in combatting arson (Figure 1-1)

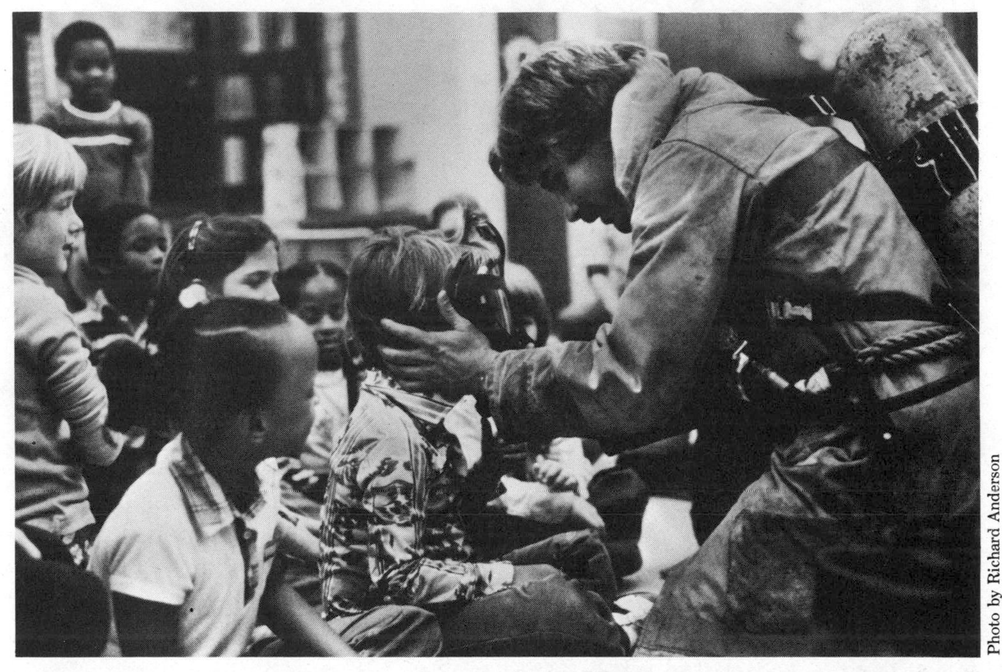

Photo by Richard Anderson

Figure 1-1. *Educational programs alert the community to the role of firefighters and the problems of arson.*

6. Special telephone numbers that citizens may use to transmit information about arson fires, and rewards for information leading to arson convictions
7. Juvenile firesetter programs, aimed at reducing the number of fires set by children, and including psychiatric and social therapy when necessary
8. Pinpointing of the structures and areas that are most likely to become the targets of arsonists, often with the aid of computerized data-processing systems
9. Surveillance of likely targets of arsonists, either electronically or by citizen, police, or fire department patrols
10. Condemning and razing or sealing of abandoned buildings.

The first of these, the immediate investigation of all fires by fire department personnel, is called *fireground investigation* and is the subject of this book. It is one of several activities, such as emergency medical care and fire-prevention programs, that fire departments have instituted to better serve their communities.

THE SCOPE OF FIREGROUND INVESTIGATION

A fireground investigation is a systematic search of the fire scene for information about a fire. Its purpose is to reconstruct the events that led to the fire—to seek out the cause of the fire. A fire investigation *is not* an arson investigation, although it will reveal cases of arson. In addition, a fire investigation can provide much information that is of use in preventing and fighting fires.

Specific Objectives

There are four primary objectives in a fireground investigation. They are:

1. Find the *point of origin* of the fire
2. Find the *heat source*
3. Determine the *reason* for the fire
4. Determine the *category* of the fire.

Point of Origin. The point of origin is the exact location at which the fire started. It is not as difficult to find as you might expect, even in a severely damaged structure. The point of origin is important in determining how the fire started and how it spread.

Heat Source. The heat source is just that—the source of the heat energy that started the fire. The most obvious heat sources are open flames and hot embers; a less obvious heat source would be solar rays focused on a combustible object by window glass, or friction in a motor bearing.

Reason. The reason for a fire is the circumstance or set of circumstances that led directly to the fire. In a sense, the reason is the method by which the heat source and the combustible fuel came together at the point of origin to start the fire.

Category. The category of a fire can only be determined after the point of origin, heat source, and reason are found. These, along with other information determined during the investigation, will usually point to one of the three possible categories—natural, accidental, or arson:

- A *natural* fire is one that takes place without human action or intervention. A fire that results when lightning strikes an unprotected structure would be categorized as natural (Figure 1-2).

Figure 1-2. *Lightning: a natural fire.*

- An *accidental* fire is one that results from human carelessness. For example, grease that is dropped on the live burner of a kitchen range (Figure 1-3), or an installation defect in electrical wiring (Figure 1-4), could lead to an accidental fire.
- An *arson* fire is one that is started deliberately and maliciously, with the intent to cause damage to property. Arson fires can be ignited by common household products (Figure 1-5).

Many textbooks use the terms *incendiary* and *arson* interchangeably. However, according to Dan J. Carpenter, Fire Administrator and Chief Fire Marshal of Mecklenburg County, North Carolina, "Incendiary means that a fire was deliberately set by a person or persons." According to this definition, an incendiary fire may or may not be an arson fire, depending on the intent of the person who set the fire. The burning of an abandoned building by fire department personnel for training purposes is incendiary but not arson. The burning of a pile of leaves by young vandals for "kicks" is incen-

Figure 1-3. *Leaving a pan of grease on a stove can lead to an accidental fire.*

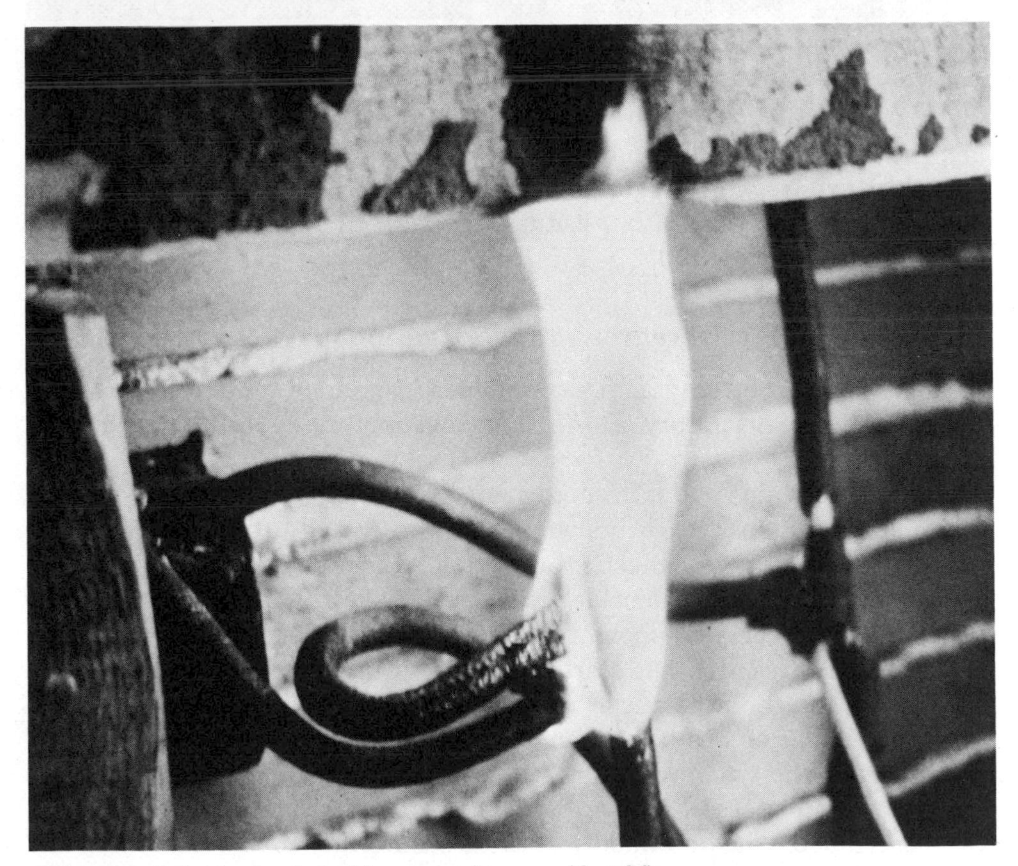

Figure 1-4. *Defect in electrical wiring can lead to an accidental fire.*

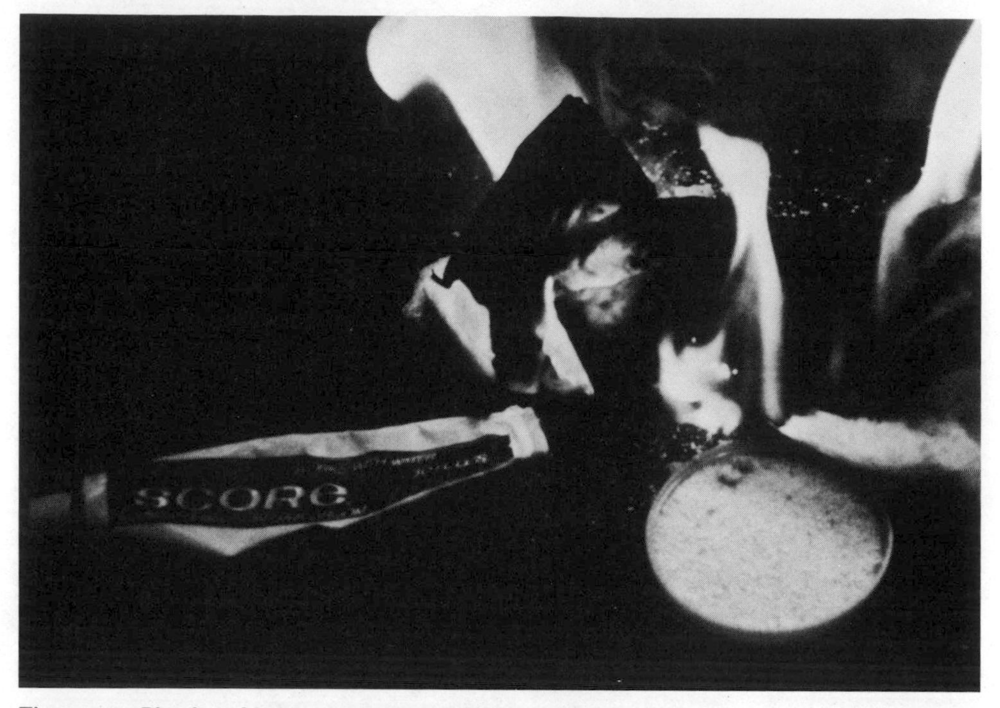

Figure 1-5. Bleach and hair cream create fuel for arson fire.

diary, but not necessarily arson, even if the fire spreads to a nearby building; the intent to cause damage to property would have to be proved.

In this text, we will categorize fires only as *natural, accidental,* or *arson.* A fire may be incendiary, but if it is not shown to be arson then it will be categorized as accidental.

CASE HISTORY: Fireground Investigation

On a hot summer afternoon, fire companies responded to a fire on the third floor of a college dormitory. The fire involved only one room and was quickly extinguished.

The smoke stains and burn marks left by the fire were heaviest near the top of a dresser, close to a window. On the dresser was an electric hair curler with a plastic casing; the plastic was completely melted. Books piled on the dresser were burned on the side facing the curler, but not on the side away from it. Plastic curtains above the dresser had caught fire and burned. It seemed likely the curler had started the fire. The investigator photographed everything and took the curler when he left the scene.

Later, the curler was taken apart and examined. The investigator found that it was of the type that could be left plugged into an electric socket. An internal thermostat, which was supposed to control the curler temperature, had failed. The unit had overheated sufficiently to melt its casing and start the fire. The investigator then was able to report the following:

Point of origin: Dresser top
Heat source: Electric hair curler
Reason: Thermostat failed, resulting in overheating
Category: Accidental.

The investigator's photos and the curler were later used as evidence in a subrogation suit arising from the incident.

Fireground Investigation

At some fires the fireground investigation can be completed in less than an hour. At others it may go on for several days. The time required for a thorough investigation depends mainly on the severity of the fire and the extent of the damage. In any case, once the four primary objectives are realized, the fireground investigation is essentially complete. What remains is for the investigators to report their findings to the appropriate agencies. If those findings include the possibility of arson, the fireground becomes a major crime scene, and a criminal (arson) investigation begins. An arson investigation proceeds under the direct supervision of a law enforcement agency such as the fire marshal's office,* the sheriff's office, or the state or local police. In some jurisdictions, fire department personnel have police powers and may conduct the arson investigation.

Fireground investigation and arson investigation differ in several ways. First, their goals are different. The objectives of a fire investigation are only to find the point of origin, heat source, reason, and category of the fire. The goal of an arson investigation is to investigate a major crime and bring the criminal to justice.

Second, the personnel are usually different. Fireground investigation is performed by firefighters who are experts in fire behavior but have little or no police training or police powers (Figure 1-6). An arson investigation is a criminal investigation performed by law enforcement personnel. (In many cases, these law enforcement personnel—including some fire marshals—have little fire training; they rely heavily on fireground investigators and other technical experts for facts concerning the fire.)

Third, fire investigation and arson investigation differ in scope. The fire investigation is more or less confined to the fireground and surrounding area. And the subject of the investigation is the fireground. An arson investigation can range far and wide, as investigators search for an arsonist and a motive. Arson investigators may have to deal with such things as the financial stability of a business or the psychological traits of a suspect, as they gather evidence for a criminal court action.

---

*In most jurisdictions, the fire marshal's office must investigate every fire in which damage exceeds a set dollar amount or where loss of life occurs.

Initial Versus Advanced Fire Investigation

The fireground investigation that we have been discussing is, in a sense, the initial investigation. It begins as soon as possible after the alarm is received, and is conducted by fire department personnel. If the fire is found to be natural or accidental, then there is no further fireground investigation; if the fire is found or suspected to be arson, then the initial investigation ends and an arson investigation begins.

In some cases, however, this initial (fireground) investigation may not reveal the answers to all the questions that arise; or, the investigators may have some doubts about the point of origin, heat source, reason, and category. If for any reason the initial investigation cannot progress to a conclusion, the investigation should be upgraded to an advanced investigation.

An advanced investigation differs from an initial investigation only in the use of technical experts, testing laboratories, and other specialized resources to resolve unanswered questions and complete the fireground investigation. A technical expert or technical investigator usually specializes in a particular area—electrical and electronic systems or flammable liquids, for example. Or, the expert may be a fire investigator from a nearby city, who happens to have the advanced knowledge and experience to solve a

Figure 1-6. *Fireground investigation determines the origin of a fire.*

particular investigative problem. As another example, an investigation may be upgraded so that investigators may use the services of a commercial laboratory, to test evidence for traces of flammable liquids.

An advanced investigation may be required when the fire has involved unusual contents, or if complicated manufacturing processes and equipment were involved. It may also be necessary if the fire and/or firefighting procedures have caused so much damage that a thorough investigation is beyond the capabilities of fireground investigators. Note, however, that an advanced investigation is still a fireground investigation; its purpose is to determine the point of origin, heat source, reason, and category of the fire. If the category is found to be arson, then again, an arson investigation is begun.

Benefits of Fire Investigation

One obvious benefit of fireground investigation is the discovery of instances of arson and the prosecution of the arsonists. Where fireground investigations are routinely conducted, arson discoveries have increased. As the investigators have become more proficient, they have contributed more information and usable evidence to fire marshals and police. There have been more arrests of arsonists, and an increased percentage of convictions. These have led to a decrease in the number of fires, with reduced loss of lives and property.

Where fires are not investigated, arsonists have little fear of discovery, and arson increases. Fire investigation is a deterrent to arson because it is so effective in the detection and prosecution of arsonists, and in removing the profitability of arson. In other words, fire investigation helps to make arson a risky enterprise, so that potential—and even "professional"—arsonists become very hesitant about setting fires.

Another, perhaps less obvious, benefit of fireground investigation is the discovery of information that is helpful in the prevention and extinguishing of fire. New materials for structures and furnishings, new industrial processes, and new products become available almost constantly. Little is known about the high-temperature behavior of many new materials and processes. Often, a problem comes to light only after involvement in a fire—when a new electrical appliance causes a fire, or a new material readily ignites or spreads flames, or a new type of construction fails quickly at an elevated temperature. Fireground investigation has uncovered such problems; many more would be discovered if all departments performed fireground investigations; and few, if any, would be discovered if fires were not investigated.

Attitudes Toward Fire Investigation

Most people and agencies involved with fire investigation have developed a positive attitude toward this activity. More and more fire departments are performing fire investigation. Fire marshals and law enforcement agencies are working more closely with investigators, to the benefit of both. Legislatures across the country are changing arson laws (or passing new ones), so that it is no longer next to impossible to convict an arsonist. The 95th Congress passed legislation that will publicize the high incidence of arson; this legislation requires the FBI to collect information on arson and to publish the number of arson cases as part of its Uniform Crime Reporting Program (FBI Crime Index).

Each of these steps is a part of the battle against arson. Every action taken by one agency reinforces the positive attitude and actions of all other agencies involved in the fight against arson. But in the last analysis, it is the attitude of the ranking officers and of firefighters that counts most. If the attitude of these officers favors fireground investigation, and if that attitude is translated into specific orders, then fire investigation duties will be performed. And if firefighters have a positive attitude, fire investigation duties will be performed thoroughly and effectively.

PERSONNEL AND TRAINING

Traditionally, the fireground is the domain of the fire service and the sole responsibility of the senior fire officer present. Where fireground investigation is performed, it too is the responsibility of the fire service. In most departments, the Chief will delegate operational responsibility for fireground investigation to an experienced officer or firefighter.

This responsibility does not, however, end with those assigned to investigative duties; it is shared, to some degree, by every officer and firefighter on the fireground. Their cooperation is essential in the investigation is to produce meaningful results.

Personnel

The personnel actually assigned to perform fireground investigation vary from department to department. In a small department, the Chief or senior officer at the fireground may perform the investigation; in other departments, a firefighter or officer may be assigned to this duty. In a larger department, an investigation team may be composed of an officer and several firefighters, or of fire officers, firefighters, and uniformed police or detectives.

Some departments are large enough to have their own photographic laboratories and testing facilities and the personnel to staff them. They can then perform advanced investigations as well as initial investigations. In most cases, though, outside assistance is required for an advanced investigation.

Actually, every firefighter on the fireground should consider himself a member of the investigation team. By committing to memory what he saw and heard—and even smelled—and what work he performed, a firefighter can contribute important information to the investigation. Fireground investigators will, of course, examine the fire scene carefully and question occupants and other witnesses. They will also question firefighters, whose training makes them excellent observers (even during complex firefighting operations) (Figure 1-7). Very often, a firefighter's recollections provide the key elements in the investigators' determination of the point of origin, heat source, reason, and category of the fire.

Training for Fireground Investigators

Fireground investigation is still a relatively new field, and training—although very important—may not be readily available. The newly assigned

Photo by Richard Anderson

Figure 1-7. *The observations of trained firefighters help investigators.*

investigator may rely on his experience as a firefighter to a great extent. A knowledge of fire science, firefighting techniques and procedures, fire behavior, and building construction is an important foundation for fireground investigation. However, this knowledge is utilized differently in investigation than in firefighting. For example, a firefighter is concerned with how effective venting operations and hoseline placement are as *firefighting tactics;* a fireground investigator considers their effect on *fire behavior* and *fire spread.* Forcible entry, too, must be considered in a new light: It is essential to be able to determine whether firefighters made the entry or if a door or window was forced by an occupant or an arsonist.

The investigator must learn a number of new skills and expand his knowledge in many areas. For example, every firefighter should have some understanding of automatic fire-detection and extinguishing systems. The investigator needs to know *as much as possible* about such systems, for he must often decide whether a malfunction was accidental or intentional. The investigator must also learn to recognize, collect, and preserve evidence; learn questioning techniques; and learn how to locate and use such records as building blueprints and fire reports. Of course, he must also learn how to make a thorough examination of the fireground, and *how to interpret the results of his investigation.*

Training for fireground investigation is never completed; it is a continuing process. This book, for instance, covers all the basics of fireground investigation, but only the basics and only what was known at the time it was written. New firefighting techniques and construction techniques, new materials, and especially new investigative techniques should be sought out and studied by both recently appointed and experienced investigators. There is always something more to be learned about the structures, people, processes, and laws involved in fire and arson. In the remainder of this section, we list some sources of formal and informal training for all investigators.

Formal Training. The need for trained fireground investigators and the benefits of investigation have prompted a number of organizations to offer fire investigation training. These include:

- International Association of Arson Investigators, Inc.
- National Fire Academy
- State fire investigation associations
- Vocational schools, community colleges, and universities
- State fire schools
- Insurance companies.

These sources may offer beginning or advanced courses in fireground investigation. They should be contacted directly concerning the type, length, cost, and availability of courses.

Informal Training. The training for many new investigators will consist only of on-the-job training combined with individual study. This learning by doing, under the guidance of an experienced investigator, may be the best way to learn fireground investigation. However, it can be extremely difficult for a new investigator, working alone, to learn by trial and error. The new investigator's training and experience as a firefighter will be of help, but they should be reinforced by additional study.

Individually and informally, both new and experienced investigators should:

- Learn as much as possible about fire behavior and building construction, for increased understanding of the effects of structural features on fire travel
- Keep abreast of newly developed firefighting techniques and fire protection systems
- Subscribe to technical journals, such as *The Fire and Arson Investigator,* to learn of new investigative techniques and concepts
- Obtain and study available texts on fire and arson investigation
- Attend seminars on arson suppression
- Confer with investigators in neighboring jurisdictions, to learn from their experiences.

Training of Firefighters

At almost every fire, the firefighters making the initial attack are the first to see the fire; the first to see occupants and bystanders at the fire scene; the first to enter the fire building; and often the first to see evidence indicating how and why the fire started. It is important that they *observe* as well as see, and that they commit their observations to memory. For this, firefighters must know what is required of them—what information is of significance in a fireground investigation. That is, they need some training in the basics of fireground investigation.

Unfortunately, firefighters are sometimes responsible for the mutilation of evidence at the fireground. In most cases, this is an unavoidable consequence of firefighting operations; however, if firefighters can recall the actions they took, investigators can often use this information to reconstruct the situation as it was prior to the fire attack. On occasion, firefighters have destroyed evidence needlessly, simply because they didn't realize it was evidence. Again, some training in the elements of fireground investigation would be of help. In fact, the more training firefighters receive, the more effective the entire fireground investigation program will be.

The Team Concept

Fireground investigation requires that a wide range of knowledge be applied to a variety of tasks in a difficult situation. The job may be too much for one investigator to do effectively, especially at a major fire. For this reason, and where department budgets allow, a number of departments have created fire investigation *teams,* consisting of two or more members.

For best results, an investigation team should consist of at least one fire officer, an experienced firefighter, and a police officer (or detective). The team should respond to fires as a unit. Firefighter members can concentrate on the fireground while police members question occupants and witnesses. This minimizes delays in the questioning process and thus results in more accurate verbal information. It also allows each team member to perform the tasks he knows best, while providing enough manpower to do the job effectively. Further, team members can help each other as necessary to reach a joint, or team, determination of the reason and category of the fire.

The inclusion of a police investigator on the team has several advantages: Police are trained investigators; they know how to question witnesses, know the law, have arrest powers, and can solicit the help of other police officers or law enforcement agencies with a minimum of delay. (Whether or not a team includes police personnel, it is essential that investigators develop a good working relationship with local and state law enforcement agencies.)

The number of personnel that can be assigned to fireground investigation depends on the department's size, manpower situation, and budget. However, there are ways to get around even a very tight budget. For example, smaller departments may solicit volunteer help from retired police officers. In many localities with volunteer departments, law enforcement personnel respond to fires for traffic-control duties; the skills of these people could be put to much better use in assisting the fire investigator.

ARSON AND THE ARSONIST

One main objective of fireground investigation is to uncover—and thereby deter—arson. We therefore end this introductory chapter with a brief discussion of arsonists and their motives.

Financial Gain

Perhaps the most common motive for arson is financial gain. The financial gain is usually the money collected on a fire insurance policy. For example, a business person may set fire to an unprofitable business to collect the insurance on the building and its contents. Or he may have a warehouse full of stock that cannot be sold, or a building that needs expensive repairs, or one that is insured for more than it is worth because the neighborhood has deteriorated. In each case, a profit can be made by collecting on an insurance policy through an undetected act of arson.

(This does not mean that fire insurance is an undesirable method of protection, but only that it can be misused for fraud. Insurance has saved, and will continue to save, honest property owners and business people from financial ruin in the event of a fire.)

One other type of person engages in arson for profit, and that is the professional "torch"—the arsonist for hire. Here, the profit is the payment received by the arsonist for setting the fire. The "amateur" arsonist—the owner who sets fire to his own home or business—may use a simple or elaborate scheme to set the fire and provide an alibi. But because he has little experience with fire and doesn't understand what it is, he generally makes mistakes that can be detected by fireground investigators. The professional torch is much more knowledgeable, and his work may be more difficult to detect. However, the professionally set fire is not a natural or accidental fire, and capable investigators can tell the difference. Moreover, a professional arsonist may use the same firesetting technique over and over again. This unique method of operation (MO), if known to police or fire marshals, immediately brands the fire as suspicious.

Criminal Cover-up

Fires are sometimes set to cover a criminal act. A murderer may use arson to destroy incriminating evidence or hide the fact that a murder has been committed. In such cases, the act of arson is usually an afterthought, poorly executed and quickly detected. However, when the fire itself is the murder weapon, the details may be carefully planned, then the crime may be difficult to uncover.

Thieves occasionally set fires to cover their crimes. The hope here is that the fire and firefighting operations will be so confusing that the theft will go undiscovered. Conversely, theft is sometimes used to cover up an act of arson.

Sensual Satisfaction

Some types of mentally disturbed individuals derive sensual satisfaction—usually sexual in nature—from setting and watching fires. They may be stimulated by the danger and fear associated with fires, or the excitement of the crowd; they may want the chance to become a hero at a fire. These pyromaniacs often will join the crowd at each fire they set, or actually work with firefighters in extinguishing the fire and rescuing victims.

These arsonists usually set fires that are easily recognized as arson. They may pile combustible materials in a hallway, or pour gasoline on furniture, and ignite it with a match. However, since this type of arsonist is not rational, there is often no pattern to their targets. This can make them difficult to apprehend.

Other firesetters with mental disorders include those who set fires because they hate the person whose property or life they are destroying, and those who "hear voices" that tell them to set fires.

Revenge

In some cases, fires are maliciously set in response to some real or imagined injustice. A person who feels he was "taken" in a business deal may set a fire to get even. Or a person who is jealous of their lover may become angry enough to throw flammable liquid at their doorway and ignite it.

Vandalism

Vandalism fires are most often set by groups of two or more juveniles—usually boys. The goal is generally excitement—kicks—rather than destruction, and the target a school or other symbol of authority. These young arsonists may also exhibit such antisocial behavior as stealing, lying, truancy, and running away from home.

Acts of Violence

Fires may be set during riots, as one of several forms of group violence. They may also be set covertly by fanatical members of otherwise peaceful groups. For example, a polluting chemical plant may become the arson target of ecological fanatics; or an unyielding employer involved in union contract negotiations could be the target of angry workers.

Racism has led to crimes of arson directed against members of minority racial and ethnic groups. Usually, the arson is not very subtle, because its goal is to create fear. However, the arsonists don't want to be identified as individuals, and they take great pains to cover their tracks.

2

INVESTIGATIVE ASPECTS OF FIRE BEHAVIOR

A fireground investigation is essentially a search for information about the fire. From the information gathered, the investigator deduces where and how the fire started and then categorizes it as natural, accidental, or arson.

For the most part, two types of techniques are used to seek information concerning the fire. One type is *questioning techniques,* used to determine what the occupants and other witnesses might know about the fires and what firefighters have observed; these are discussed in Chapter 9. The other type may be called *examination techniques.* The investigator uses these techniques to examine the fire structure and surrounding area for signs—or clues—as to what took place.

The fire itself is responsible for the majority of these clues. That is, as the fire travels, it leaves telltale signs indicating where it started and, often, how it started. To find and interpret these signs, the investigator must apply his knowledge of fire science and fire behavior. This is the first of two chapters that explain how that is done.

IGNITION AND BURNING

As a quick review, we note that three things are necessary for fire to occur. These are fuel, oxygen, and heat (a source of ignition). There must be sufficient heat to vaporize some of the fuel (if it is a solid or liquid) and to ignite the vapor after it mixes with oxygen. For sustained burning, the fire must produce enough heat to vaporize more fuel, which then must mix with more oxygen and burn; this in turn must produce more heat and fuel vapor, and so on. This process is often called the *chain reaction*. Fuel, oxygen (usually from the surrounding air), heat, and chain reactions form the faces of the so-called *fire tetrahedron.* When one of these elements is removed, the fire dies out; this is the basis of every method of extinguishment.

Burning is thus a chemical and physical process that proceeds according to certain laws of chemistry and physics. In exactly the same circumstances, two fires will burn in exactly the same way. (If your experience tells you no two structural fires are alike, that is because, in reality, no two sets of circumstances are alike. Even minor differences in the fire's location, the fire building, its contents, the wind speed and direction, firefighting operations, and many other factors will cause differences in fire behavior.) Once a fire starts, the circumstances dictate its behavior. This means that a trained investigator can reconstruct where and why a fire started if he can determine (1) how the fire behaved and (2) the circumstances in which it occurred. These are important goals of the examination phase of fireground investigation.

Oxygen

Oxygen is usually present in the surrounding air in sufficient quantities to support combustion. Thus, it is rarely a factor in a fireground investigation. However, an arsonist may open doors and windows to provide a draft that will accelerate the fire. This situation should come to light during the investigator's examination of the fire structure (Chapter 6) and questioning of firefighters at the scene (Chapter 5).

Fuel

Most residential, commercial, and industrial structures contain a variety of fuels sufficient to support heavy combustion. The contents and interior finishing materials (walls, floors, ceilings, and their coatings) found in most buildings will produce flammable vapors when heated. Some of these fuels vaporize slowly and thus burn slowly; others vaporize and burn very rapidly. But every bit of fuel can add to and sustain the fire.

The combined flammable contents and flammable finishing materials within a structure are referred to as its *fire load.* Normally, the fire load (especially furnishings, stock, raw and finished materials, etc.) is distributed throughout the occupancy in some logical pattern. The investigator should become suspicious when this pattern is disturbed. For example, excelsior and other packing materials would be normal in a warehouse, but highly suspect in an adjoining office area. Fuels are discussed in more detail in Chapter 3.

Flammable and Combustible Liquid Fuels

Flammable and combustible liquid fuels are discussed in several places in this book because they are used extensively by arsonists. They are easy to obtain and transport; they can be quickly splashed about the interior or exterior of an arson target; they are easy to ignite; and they burn quickly, accelerating and spreading the fire.

Among the common flammable and combustible liquid fuels (or *accelerants*) are gasoline, kerosene, lighter fluid, and cleaning fluid. Gasoline, for example, releases flammable vapors at normal room temperature; the vapors, if not confined, mix with air and need only a source of ignition. Once ignited, the gasoline produces about 2½ times more heat, pound for pound, than wood. And it produces this heat several times faster than wood. The result is an intense fire that quickly travels to other available fuels.

Ignition (Heat) Source

For a fire to start, heat must somehow be applied to the fuel in the presence of oxygen. Most occupancies contain a number of heat sources. Some are quite obvious, like matches or a kitchen range; others, including certain combinations of household chemicals, are not so obvious. Heat sources are the subject of Chapter 4.

One of the basic objectives of fireground investigation is to find the ignition source. The investigator must also find the point of ignition. When he does, he is well on the way to completing the investigation. If no connection can be found, then something is obviously wrong.

Products of Combustion

Four important combustion products are given off by a fire:

- Heat, generated by the chemical reaction called burning
- Smoke, which is made up of visible particles of unburned fuel
- Gases, which are invisible products of complete or incomplete burning
- Flame, made up of burning fuel vapor along with glowing hot fuel particles.

These combustion products do not remain at the seat of the fire. The heat, smoke, and gases rise away from the fire; the flames lick upward and outward, as they seek fuel. This movement has two consequences. First, the fire will spread to other areas in a manner that depends on the structure and its contents. The route that the fire will take can be predicted from the natural laws of chemistry and physics and, particularly, from a knowledge of heat transfer methods. Second, as the fire travels, the heat, smoke, and flames will leave their marks on walls, ceilings, floors, and contents. The patterns of these marks can be used by the investigator to trace the path of the fire back to the point of origin.

Stages of Fire

Most fires begin small and increase in size and intensity if fuel and oxygen are available. Within a structure, the oxygen supply may become depleted as the fire grows. Then the fire will proceed through three distinct stages.

Stage one is the initial development of the fire. Oxygen is in plentiful supply, so burning is relatively complete. As a result, there is rapid burning, with vigorous flames and minimal smoke and heat. The temperature is 100 to 800°F (37°C – 426°C).

As oxygen is consumed and the supply of oxygen is lowered, the fire enters *stage two*. In this stage, flame and heat production increase, and the temperature rises to 800 to 1000°F (426°C – 537°C). The reduced oxygen level results in incomplete burning, so that more smoke and gases are produced.

In a closed building, the fire can reach *stage three,* in which the oxygen supply is reduced drastically. The fire recedes to its point of origin and smolders, since it cannot propagate without sufficient oxygen. Heat production is very high, with temperatures of 1000 to 2000°F (537°C – 1093°C). The incomplete combustion produces heavy smoke that leaves a dark brown stain; this stain can be seen on windows from outside the fire building. The smoke is trapped within the closed building, along with very hot, combustible gases (mainly carbon monoxide). These gases are heated above their ignition temperature; they lack only oxygen for ignition. If air is introduced into the building before the gases are vented from above, they can ignite with explosive force. This sudden ignition of hot, unburned gases is termed *backdraft.*

CASE HISTORY: Oxygen Depletion

The County Sheriff was asked to investigate a house on the edge of a Wisconsin town when none of its occupants had appeared in town for two days. The house was a sturdy old two-story frame building, heated entirely by two wood stoves. The temperature in the area had been below freezing for close to three weeks.

The sheriff noted that the front windows of the house, covered with plastic for insulation, were heavily stained with smoke residue. The rear windows, also covered, were unstained. He could see evidence of burning in the dining room, around the dining room stove.

The sheriff forced the rear door and entered the house to conduct a search. He found the entire family dead—the husband at the top of the stairway, the wife in the upstairs hallway, and the three children in bed. The county coroner ruled that all five victims had died by asphyxiation.

The fire investigator noted that the two wood-burning stoves were located on the first floor. Heat was allowed to rise to the second floor bedrooms through gratings in the first floor ceilings. To conserve heat, all windows had been covered with plastic, doors had been weather stripped, and cracks had been sealed with caulking. The house was, essentially, a sealed container.

The investigation established the following sequence of events: A fire originated in an upholstered chair that was left too close to the dining room stove. It consumed part of a stack of books, magazines, and newspapers, scorched the wallpaper, but did not extend any further.

As it burned, the fire taxed the limited oxygen supply in the dining room to the point at which open flame production could not be sustained. After a brief period of open burning, the fire receded to the smoldering stage, consuming the chair and producing deadly combustion by-products. The heat, smoke, and gases moved up through the ceiling heat registers to the second floor, where the occupants were sleeping. The toxic gases accumulated in sufficient quantities to result in the deaths of the entire family.

The investigator determined that the point of origin was the upholstered chair; the heat source was the stove (and ignition was probably through radiation); the reason was the placement of combustible furniture too close to the heating device. The fire was categorized as accidental.

FIRE SPREAD

A fire will spread away from the point of origin if sufficient fuel and oxygen are available. This fire spread occurs through a single mechanism—the transfer of heat to available fuels. From the path of the fire and the patterns it leaves, the investigator can determine whether the fire spread naturally, or whether it was helped along by an arsonist.

Methods of Heat Transfer

Heat is transferred from a fire to new fuels by four methods: convection, conduction, radiation, and direct contact.

Convection. Convection is the transfer of heat by the motion of hot smoke, gases, air, and particles. Hot smoke and gases tend to rise away from the seat of the fire. The air near the fire is heated by the fire, and it also rises (because it is lighter than cooler air). The rising air, smoke, and gases carry hot particles and embers away from the fire. As these heated gases and solids move away, cooler air moves in toward the fire. This sets up currents that accelerate the convection process and are, in turn, accelerated by the increasing rate of combustion.

If the vertical movement of the hot convection currents is blocked, say by a ceiling, the gases and particles will move horizontally, and through any available openings. They will again move vertically when that is possible (Figure 2-1).

The moving gases and particles leave a distinct stain pattern on walls, ceilings, and contents. Because these combustion products move up and away from the fire, the wall pattern tends to resemble a wide V, with its lowest point at the point of origin of the fire. The lower parts of walls and contents may remain untouched, while the ceiling and upper walls are heavily stained by smoke and blistered or charred.

Hot particles and embers can be carried for some distance by convected smoke and gases. If they then drop onto combustible materials, they can start secondary fires. Such a "remotely ignited" fire will grow in size if it is not extinguished quickly. It will then produce a point of origin very much like the original. This situation poses a problem, because two or more points of origin without some logical connection may also indicate arson. The fireground investigator must determine why separate fires occurred and whether or not one of them led to the others through convection.

Conduction. Conduction is the transfer of heat through a solid material. Metal objects, such as beams, nails, pipes and wires, are excellent conductors of heat. Many people have discovered this the hard way—by grabbing

Figure 2-1. *Ceiling blocks vertical movement of hot convection currents.*

a metal object at one end after heat was applied only at the other end. Similarly, heat can be conducted from an involved room into an adjoining room by, say, a cast iron pipe. Combustibles in the adjoining room may be ignited, even though the two rooms seem isolated from each other.

Wood is not a good conductor of heat. However, wood that is in contact with a heated surface or object can become pyrolyzed (that is, decomposed to vapor) over a period of time. The heat of the surface or object is then sufficient to ignite the pyrolyzed wood, producing open flames. This pyrolyzation and ignition may also occur much more quickly when heat is conducted to a wooden object from a fire by a metal beam or pipe.

Fire spread by conduction leaves no pattern for the investigator to trace. However, the conductor itself should lead the investigator back to the point of origin of the original fire (Figure 2-2).

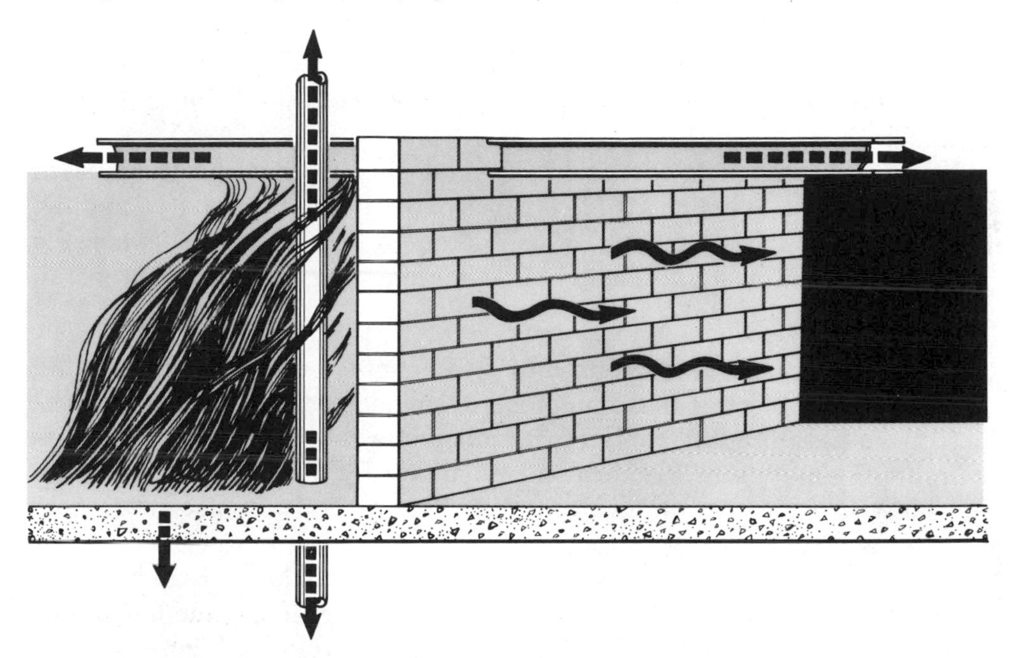

Figure 2-2. *Heat conduction causes other areas to ignite.*

Radiation. Heat radiation is the transfer of heat as invisible waves across a space, in the same way light is transmitted. Radiant heat waves travel in straight lines in all directions. They will move through air; are not affected by winds; and will penetrate transparent and translucent surfaces, including glass and water. Radiant heat waves are absorbed by any solid, opaque body that they contact. The absorbed heat produces flammable vapors that mix with surrounding air and are then ignited by additional radiant heat waves.

Radiant heat moving through windows has caused ignition in exposures (Figure 2-3). Heat radiated from flames traveling in ceiling air-conditioning ducts has been known to ignite papers lying on an office desk. Material as

Figure 2-3. Radiant heat waves travel through translucent materials and can cause ignition away from the original fire.

far as 100 feet (304 decimeters) from a large fire has been ignited by radiation.

Ignition by radiant heat would seem to pose a problem for the investigator, because there is no visible contact between the heat source and the fuel. However, there must be an unobstructed line of sight between the source of radiant heat and the ignited material; otherwise, ignition by radiation cannot occur.

Direct Contact. Heat is transferred by direct contact when a flame or ember touches some object. If the contact is maintained for a sufficient time, the object will be ignited. We noted that embers may be carried by convection currents. Flames from matches cause ignition by direct contact. Flames reaching curtains from an involved chair (Figure 2-4) will also cause heat transfer and fire travel by direct contact.

The path of fire travel will usually indicate when fire has spread by direct flame contact. Some reconstruction may be necessary if intermediate flame carriers have been destroyed. For example, the curtains in Figure 2-4 may have transmitted flames to the wall and ceiling before falling. Although fire spread from the chair to the wall was by direct contact, the "link" between the two is no longer in place. The investigator should be able to reconstruct what happened from the burn on the floor and the fact that the curtains are missing.

Avenues of Fire Travel

Vertical. Fire travels via the transfer of heat, and heated air, smoke, and gases tend to move upward. Thus, a fire will spread vertically when construction features allow it to do so. Open stairways, elevator shafts, pipe shafts, and spaces between exterior and interior walls provide vertical pathways for the upward movements of hot combustion products (Figure 2-5).

Flammable materials (including contents) in or near the pathway will become involved if the combustion products are hot enough to cause vaporization of these fuels and then ignite the vapors.

When the vertical movement of heated combustion products is obstructed, say, by a ceiling, they fan out in all directions. They travel laterally along the ceiling until they encounter another obstacle, such as a wall. If there are no openings in the wall, the combustion products accumulate until they are forced downward along the wall. This horizontal and downward motion is called *mushrooming* (Figure 2-6). It leaves a recognizable pattern of smoke stains on the walls. If the combustion products are hot enough to ignite the ceiling and wall materials, there will also be evidence of burning.

Figure 2-4. *Direct flame contact.*

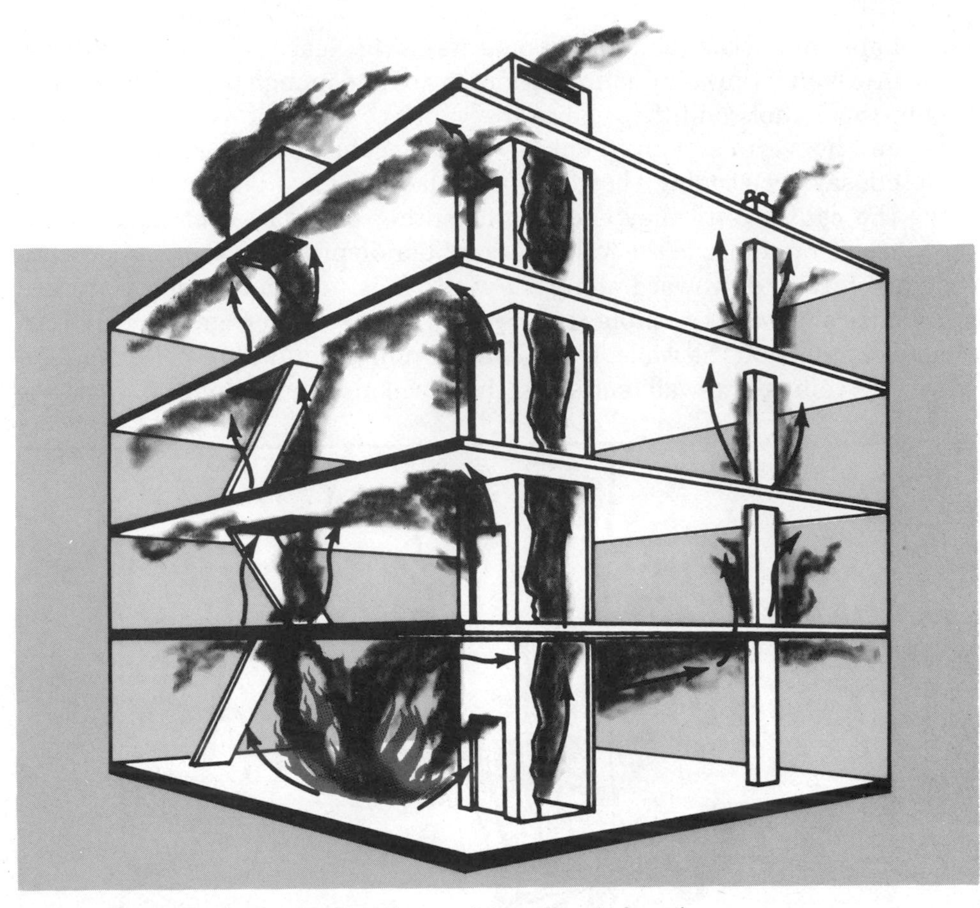

Figure 2-5. *Combustion products travel upward through vertical openings.*

Horizontal. If mushrooming combustion products encounter an opening, they will move through it to an uninvolved area. They then move upward if possible, or horizontally if no upward path is available. Again, combustible materials along the path are ignited, and the fire spreads. In this way, fire can travel horizontally through a long hallway at ceiling level, with almost no burning on the walls.

Large, open areas, such as the interiors of churches and supermarkets, allow fire to spread very rapidly at or near ceiling level. Very hot gases collect in the upper reaches of the building. When the proper conditions are reached, they ignite. Fire flashes across the area very quickly, and can involve the entire structure within minutes. It is believed that this is what happened in the unprotected, open casino of the MGM Grand Hotel in Las Vegas in November 1980.

Downward. Fire travels downward mainly when burning material drops from an upper area to a lower one. This is a common occurrence in balloon-frame buildings (discussed in Chapter 3); embers from an attic fire can drop all the way to foundation level and ignite combustibles at that level.

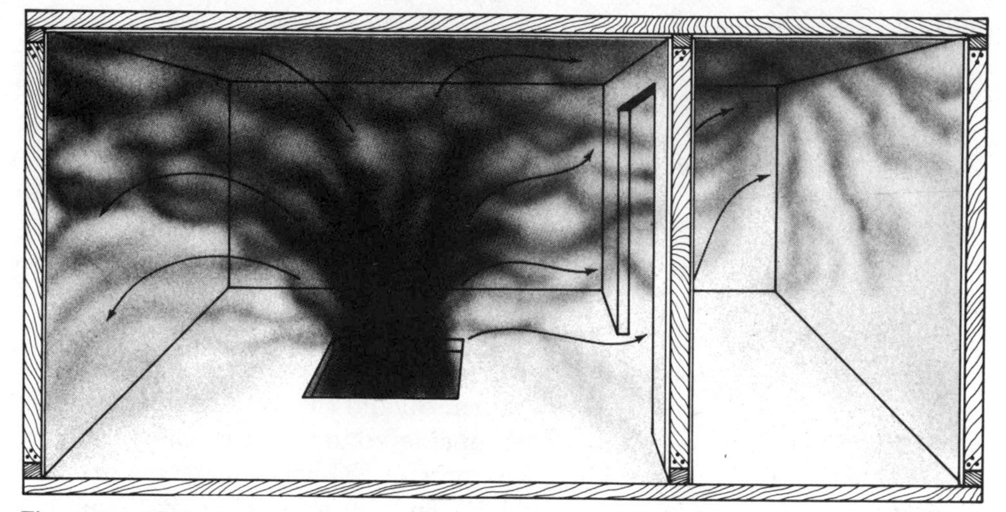

Figure 2-6. *Mushrooming pattern.*

Fire can also be carried downward by wall coverings such as varnish, paint, wallpaper, and flammable paneling. However, this is a very slow process that usually accounts for only minor fire spread. The blower of a forced-air heating or cooling system can push fire down the ducting. In fact, any vertical shaft provides an opening through which burning embers can drop downward and ignite another fire.

Fire will follow flammable liquid that is poured down a sloping surface. For example, if gasoline is poured on a stairway and ignited at the top stair, the fire will travel down the stairs as it consumes the vapors. Eventually, the fire will travel back up the stairs as the flammable liquid is consumed and nearby fuels start to vaporize.

EXPLOSIONS

There are four types of explosions: deflagration, pressure release, decomposition, and detonation. Deflagration is the only type directly associated with fire. The other three can occur with or without fire, but the damage they do can initiate a fire.

Deflagration

Deflagration is a very rapid burning that produces intense heat and often brilliant light. It can occur when a combustible gas or a dust becomes mixed with air in particular proportions and is then ignited. Under many conditions, the mixture burns so fast that it produces an explosion.

Gaseous Fuels. Natural gases, the vapors of liquid petroleum and flammable liquids, and the flammable gases produced by incomplete combustion are all subject to deflagration. These gases and vapors can accumulate within a structure and mix with the air in that structure. If they are not dispersed by air currents, they can remain in place for hours, and can be ignited by an ignition source as simple as the flipping of an electric switch. However, the proportion of gas or vapor in the air must be within certain limits or the mixture will not ignite.

To explain "explosive" or flammable range, carbon monoxide (CO) will be used as an example. The explosive or flammable range of CO is 12.5% to 74.0% in an air mixture. This means if the atmosphere within a structure contains 12.5% CO or more, it can burn or explode. Should the percentage be less than 12.5%, such as 11.0%, it is considered to be in the *lower explosive limits* (LEL), or too lean to burn. When the percentage of the CO and air mixture is 74.0% or less, it will burn. Should the percentage of CO be more than 74.0%, such as 76.0%, it is considered to be in the *upper explosive limits* (UEL), or too rich to burn. Mixtures of CO with air between 12.5% and 74.0% are between the LEL and UEL and will ignite when they contact a sufficient heat source.

When the percentage of vapor or gas in a vapor-air mixutre is between the LEL and the UEL, that mixture will ignite and deflagration may take place. Table 2-1 shows the explosive limits for several substances. Note the wide explosive and flammable range of carbon monoxide, which is present at every fire.

Table 2-1. Lower and upper explosive limits (percentage by volume in air)

Substance	*LEL, % in air*	*UEL, % in air*
Kerosene	0.7	5.0
Gasoline	1.4	7.6
Propane	2.2	9.5
Natural gas	3.8	13.0
Carbon monoxide	12.5	74.0

If firefighters or investigators are at the scene during a deflagration explosion, they can often tell whether the explosive mixture was richer or leaner in fuel. A richer mixture—in the top half of the explosive range—creates a slower, longer-burning fire with little explosive force. It produces much heat, causing the ignition of most combustibles in the area. Ignition of a richer mixture produces a "whoosh" sound.

A leaner mixture—in the bottom half of the explosive range—creates a fast flash fire with great explosive force. It does not produce much heat, and there is only limited ignition of nearby combustibles. The explosion produces a loud bang.

These differences are illustrated in Figure 2-7 for carbon monoxide. The closer the mixture is to one of the explosive limits, the more distinct these upper-lower differences will be.

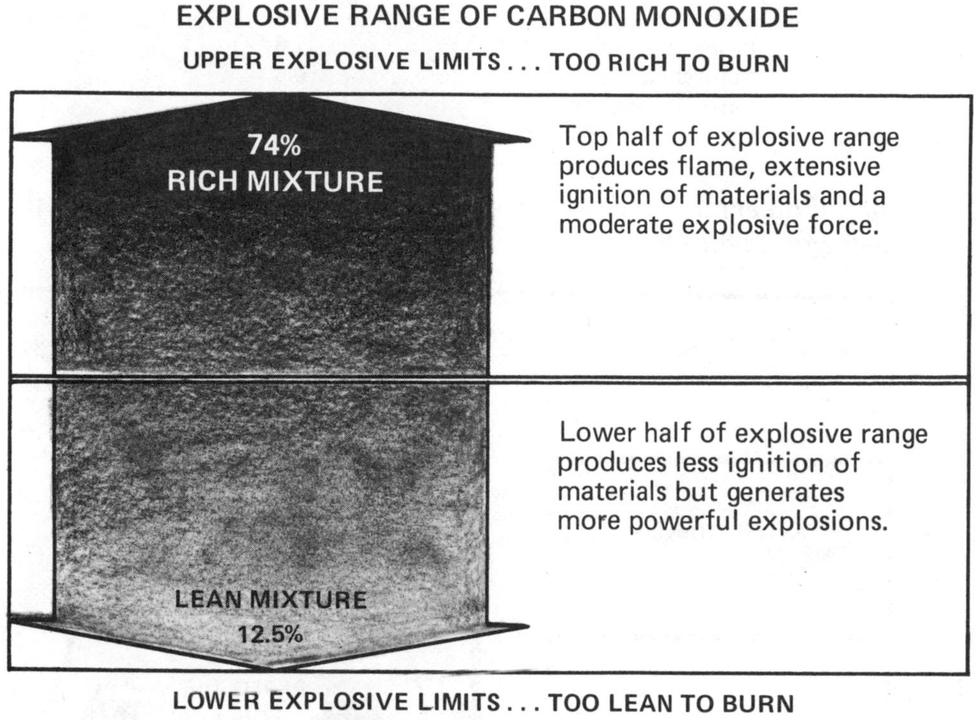

EXPLOSIVE RANGE OF CARBON MONOXIDE

UPPER EXPLOSIVE LIMITS . . . TOO RICH TO BURN

74%
RICH MIXTURE

Top half of explosive range produces flame, extensive ignition of materials and a moderate explosive force.

Lower half of explosive range produces less ignition of materials but generates more powerful explosions.

LEAN MIXTURE
12.5%

LOWER EXPLOSIVE LIMITS . . . TOO LEAN TO BURN

Figure 2-7. *Explosive range for carbon monoxide.*

The ignition of a vapor-air mixture, whether in the top half of the explosive range or the bottom half, leaves burn patterns on the walls of the structure. These burn patterns (or heat marks) are distributed rather evenly around the room, because vapors accumulate evenly within a closed space. The slow-burning fire caused by the rich mixture produces deep burning; the fast-burning fire caused by the lean mixture only scorches exposed surfaces.

Vapor Density. The density of a vapor is the ratio of its weight (per unit volume) to the weight of dry air. A gas or vapor that is heavier than air has a vapor density greater than 1. A gas or vapor that is lighter than air has a density less than 1. Table 2-2 shows the vapor densities of several substances. Propane, for example, is heavier than air, whereas natural gas is lighter than air.

A vapor that is heavier than air will drop down and collect at the lowest point in the space it occupies—near the floor in a closed room or in the basement and lower stories if they are accessible. When a heavier-than-air vapor or gas is ignited, burning is heaviest at the lower parts of the room or

Table 2-2. Vapor densities (dry air = 1)

Substance	*Vapor density*
Gasoline	3-4
Propane	1.6
Natural gas	0.6
Carbon monoxide	1.0

structure. If an explosion occurs, the lower part of the structure shows the most damage (Figure 2-8).

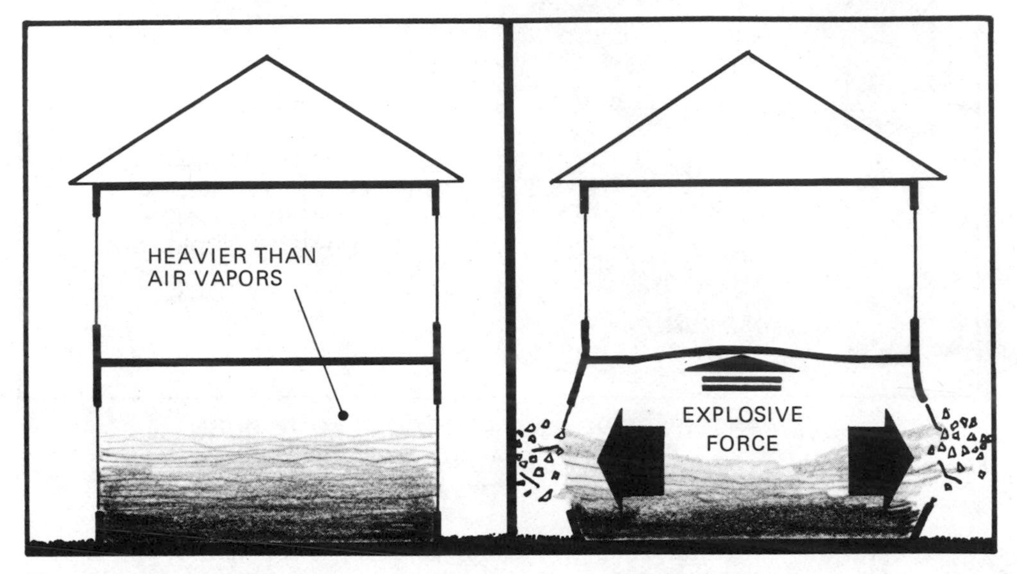

Figure 2-8. *Heavier-than-air vapors explode causing damage to lower part of a structure.*

A lighter-than-air vapor will rise and collect in the upper part of a room or a structure. If it ignites or explodes, burning or damage is heaviest near the ceiling or at the upper stories (Figure 2-9).

The differences between lighter-than-air and heavier-than-air vapor-explosion damage can be of help in fireground investigation. This is illustrated by the following case history.

CASE HISTORY: Gas Explosion

Fire companies responding to an explosion and fire alarm at a one-family residence found that the detached garage had been blown apart. Burning debris was scattered about the area. The garage's rear wall had housed a gas meter; the meter was ruptured, and gas was burning and feeding the fire. A male resident was lying on the driveway, seriously injured. He was

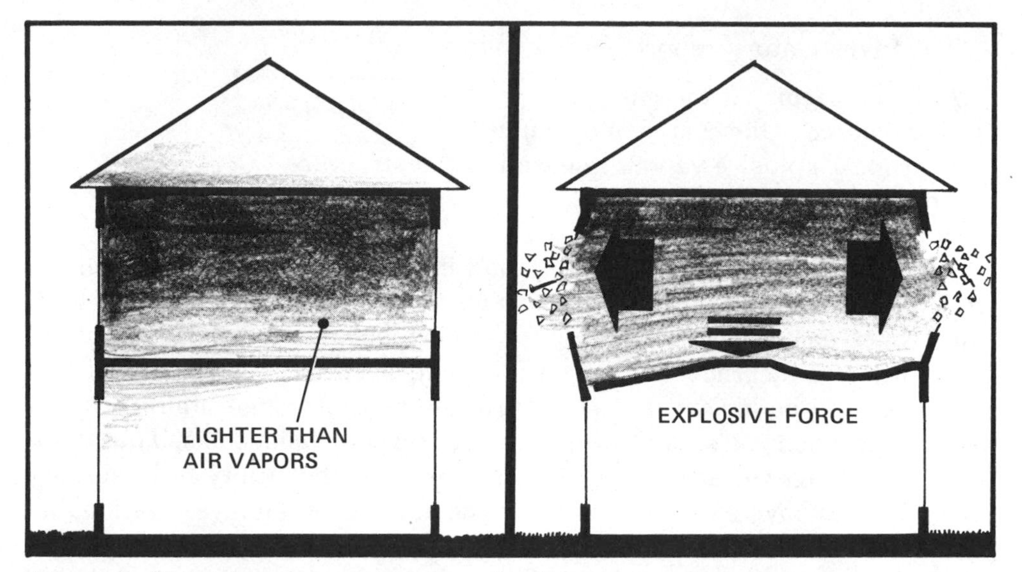

Figure 2-9. *Lighter-than-air vapors explode causing damage to upper part of a structure.*

given emergency care and transported to a hospital. The fire was quickly extinguished, with no fire extension beyond the garage.

An investigation was started immediately. The investigator had difficulty in determining which occurred first: Did a gas leak cause the explosion and fire, or did a fire cause the gas leak and explosion? Technical assistance was obtained from an insurance company specialist and a gas company investigator. It was agreed that the bottom sections of the garage had been blown out, and the top sections and the roof had fallen in. This would indicate a heavier-than-air vapor. Natural gas, which is lighter than air, would have congregated in the upper part of the structure, creating more damage in that area. In addition, inspection of the broken gas line indicated that it had been forced apart at a connection. The threads at this connection were stripped.

The investigators also found a propane tank under a section of the collapsed roof. The valve assembly was missing from the tank but was found in the debris. Because propane is heavier than air, the investigators felt certain that a propane explosion had caused the gas leak and the fire. However, they could not reconstruct the exact sequence of events.

When the resident was well enough to be questioned, he gave the following account: The tank's valve assembly was leaking slightly, and he took the tank into the garage to work on it. As he turned the valve nut, the gas leak increased. Although worried, he gave it one more turn. The valve assembly came completely off, and propane gas came whistling out of the opening. He ran out of the garage and was closing the door when the explosion occurred. Further questioning revealed that he had turned the valve nut clockwise, when in fact it must be turned counterclockwise for tightening. In addition, he had been using a small, floor-model gas heater for warmth.

The investigators' report read as follows:

Point of origin: At the gas heater
Heat source: Gas heater pilot flame
Reason: Explosive vapors released in the structure
Category: Accidental.

This case history illustrates a second important point: Do not jump to conclusions. It would have been easy for the original investigator to label the broken gas pipe as the reason for the explosion and fire. However, gas company and insurance investigators would eventually have proved that conclusion to be incorrect. Take your time; be thorough; obtain technical assistance when you are not sure. Guesses and conclusions based on superficial knowledge will usually backfire. You may feel that utility and insurance investigators have a vested interest in the outcome of an investigation, but in most instances they try to establish the truth. And if you know your job, you will be in an excellent position to assess their conclusions.

Dust Explosions. The dust of practically every type of material—wood, plastics, hay, grains, metals, and even household dust—can be extremely explosive. Fine particles of the dust must first be suspended in air within a structure. Then the temperature of at least some of the particles must be raised to the ignition temperature. Because they have so much exposed surface area, the particles ignite and burn extremely rapidly, causing a destructive explosion. The initial explosion can disperse more particles, causing another explosion. A chain of explosions can occur if there is a sufficient supply of dust particles to be dispersed and ignited.

Dust explosions occur mainly in silos and where industrial woodworking and metal-grinding operations are conducted. They occur in dwellings only rarely. In most cases, they create fairly even burn or heat patterns on the walls and ceiling of the involved space.

Pressure Release

A pressure-release explosion results when a container or vessel ruptures because of the uncontrolled buildup of internal pressure. The contents and chunks of the container may be scattered over a wide area. Fire may or may not result, depending on the contents. Containers that are subject to pressure-release explosions range from large, fixed flammable liquid or liquefied gas storage tanks to small, hand-held aerosol cans. Residential or industrial steam boilers can also become involved in this type of explosion.

The tanks used to store or transport flammable liquids such as gasoline or liquefied natural gas are equipped with safety devices known as pressure-relief valves or safety vents. These devices are designed to open if necessary to relieve excess pressure within the tank—for example, when radiant heat from the sun causes vapors to form in the tank. However, when the tank is subjected to extreme temperatures, as from a nearby fire, vapors are produced faster than the safety vent can release them. If the tank is not cooled to reduce the pressure, it will explode.

When flames impinge directly on the part of a tank container liquefied gas above the liquid level, the metal will soften and stretch. Inside the tank, additional liquid will vaporize. The excess internal pressure will cause the metal to fail, and again a pressure-release explosion will result. All the liquefied gas in the vessel will vaporize and ignite at once, producing a tremendous fireball. This type of explosion is known as a *boiling liquid expanding vapor explosion,* or BLEVE.

A steam boiler or other vessel can build up dangerously high internal pressure if its relief valve fails to operate normally. Such a failure may result from corrosion, dirt, salts (mainly in boilers), or even animal nests in or on the valve, combined with lack of proper maintenance. Failures may also be produced deliberately by wiring and wedging the relief valve in the closed position.

Decomposition

Decomposition explosions result from the fast decomposition of certain materials, including black and smokeless gunpowder, nitroglycerine, fertilizers such as ammonium nitrate, and dynamite. Some gases, such as hydrogen and acetylene, are also subject to decomposition explosions. These materials explode when exposed to heat or shock waves from other explosions, or even when struck a heavy blow. The result of such an explosion is extensive damage throughout the structure, with mainly jagged-edged debris. Most of the contents are blown away from the original positions. There is also a lack of deep burn patterns.

Detonation

Detonations are sometimes distinguished from deflagrations by the speed of the reactions: A deflagration proceeds at a speed below the speed of sound, whereas a detonation takes place faster than the speed of sound. However, we will use the word detonation to mean an explosion caused by a bomb.

The use of a bomb implies the intent to destroy or kill. A bomb explosion is therefore a criminal act. The explosion site should immediately be designated as a crime scene. If fire is not involved, the investigation is strictly a police matter. If fire preceded or resulted from the explosion, the investigation should be conducted jointly by the fire marshal's office and police. Any information or evidence obtained by a fire investigator before the arrival of police should be made available to them.

Caution. A building damaged by a bomb may be weakened structurally. Other undetonated bombs may still be in the structure. These two possibilities make it necessary that fire department activities be conducted with *very special care.*

Investigating an Explosion Site

The force of an explosion may collapse part or all of the structure and destroy much of the physical evidence. It may also push fire in all directions within the building. In spite of this, the investigator must try to discover what type of explosion occurred and why it happened. If the explosion caused a fire, burn patterns will be available for examination. If a fire caused the explosion, the burn patterns will be disrupted; determination of the cause and category of the fire may range from difficult to impossible.

The discussion of explosions in this chapter provides some basis for the investigation of an explosion site. Even rubble may provide clues as to:

1. Whether burning was heavy (deep) or superficial
2. Whether burning took place high up on walls or near the floor
3. Whether the upper or lower parts of the structure were damaged most
4. Whether the debris is unusually jagged and broken
5. The degree of disorder and damage.

This information can be used to focus on likely causes of the explosion and/or fire. For example, heavy burning high up on the walls, with damage mainly to the upper part of the structure, would indicate that the cause was a fairly rich mixture of a lighter-than-air vapor, such as natural gas. Additional help may be obtained from the U.S. Bureau of Alcohol, Tobacco, and Firearms, the national agency most concerned with the investigation of explosions, bombs, and so forth.

Again we note that investigating an explosion site can be very dangerous. The building may have been weakened, and contents may be unstable; broken utility lines can present a hazard; sheared off water pipes can fill a basement deep enough to drown in; and additional explosions are a constant possibility. Caution is essential.

3

FIREGROUND INFLUENCES ON FIRE BEHAVIOR: POINT OF ORIGIN

In the last chapter we discussed fire behavior in general terms. In this chapter we focus on details: how particular structural features, contents, and firefighting tactics affect the path that the fire takes as it travels through the fire building. The purpose of these two chapters is to provide the background the investigator needs to accomplish the first objective of fireground investigation, that is, to trace the path of the fire, from areas of least damage back to its point of origin. Since this takes place after the fact—after the fire has done the damage—it requires a good deal of knowledge, skill, patience, and concentration.

BUILDING CONSTRUCTION

Construction practices affect the fire travel mainly in two ways. The type of construction, the interior space configuration, and structural features such as elevator shafts affect the *path* of the fire as it spreads. The exterior and interior building materials affect the *availability of fuel* for fire spread. These factors, in combination with the contents of the building, may also determine the method or methods of fire travel.

Types of Construction

Nowadays, no two buildings are constructed in exactly the same way. In new buildings and in renovations, architects seem to strive for some difference in look or construction or both. Nevertheless, there are six basic types of construction: Balloon frame, platform frame, post and frame, plank and beam, ordinary, and heavy timber. Each type tends to limit or promote fire travel in its own way.

Balloon Frame Construction. Balloon frame construction is no longer used by builders. However, many existing two- and three-story homes in older residential districts were built by this method. The framing is of wood. The studs run in a continuous line from foundation to roof. Thus, the walls and studs form uninterrupted vertical shafts that extend from the basement to the eaves.

If the basement becomes involved with fire, these vertical shafts will act as chimnies to pull the fire toward the attic (Figure 3-1). If a fire above the basement level penetrates a wall, it too will be pulled toward the attic. In addition, burning materials may drop between the studs to the basement. It is not uncommon, therefore, for fires to burn only on the lowest and highest levels of a balloon frame structure. However, every level of the structure may become involved through normal (nonarson) fire travel.

A hose stream directed into the space between studs can complicate fireground investigation. Hot gases and/or fire pushed ahead of the stream will leave abnormal patterns. If the fire is divided and pushed in two directions by the stream, it may appear that separate fires have occurred. This, in turn, might lead investigators to the false assumption of arson. However, careful consideration of all the evidence, including firefighting operations and building construction, should allow investigators to categorize the fire correctly.

Platform Frame Construction. In platform frame construction, a platform is built for each story; the story is then framed in on this platform. Each platform serves as a fire stop, throughout the building. Vertical channels between studs are no more than one story high (Figure 3-2).

Fire spread in platform frame structures should be limited to the space in which the fire originated. However, holes must be cut in the platforms to pass pipes and wiring from floor to floor. If these holes are not tightly packed with fire-resistant materials, they can allow the vertical spread of fire. Concealed spaces within walls (soffits), such as the spaces above kitchen cabinets, can also become avenues of fire travel.

Post and Frame Construction. Post and frame construction provides large, open, unobstructed spaces. For this reason, it is used for such structures as churches, supermarkets, barns, and arenas. Vertical posts are placed around the perimeter of the structure; large beams or trusses are supported at their ends by the tops of the posts. The posts and beams form the frame of the structure, and the roof and walls are constructed on this frame (Figure 3-3). In most modern construction, the posts and beams are made of metal; trusses are usually wooden.

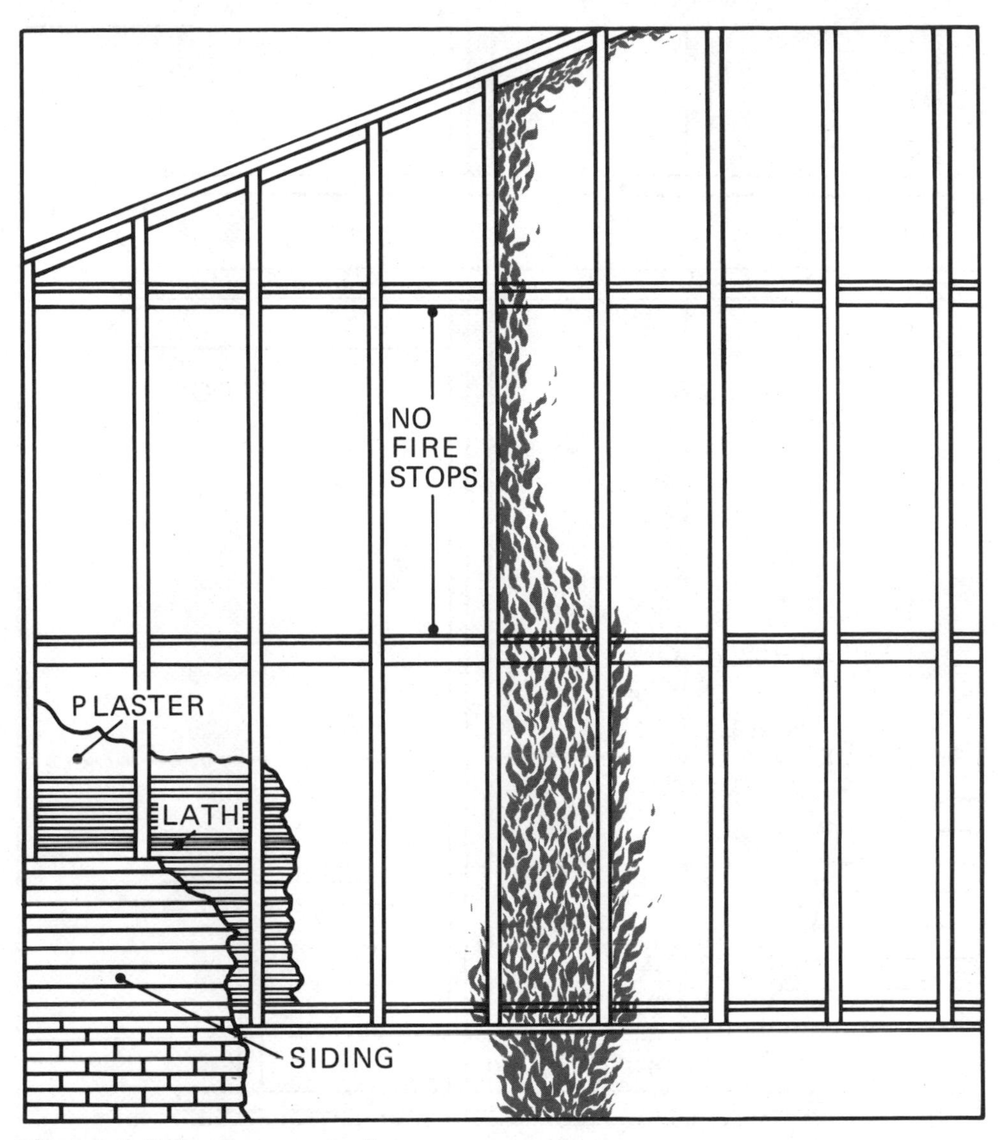

Figure 3-1. *Balloon frame construction.*

The large, open space within a post and frame structure produces a fast-moving fire. The lack of obstructions allows mushrooming and the lateral spread of hot combustion products. The structure can become completely involved within minutes of ignition. The tremendous amount of heat generated by the fire can weaken metal beams to the point of collapse.

Such a fire is very difficult to control. Unless fire companies are able to attack the fire during its early stages, damage is severe. Investigation then is difficult, but not impossible; even fires that result in the collapse of the structure can be successfully investigated. In some cases, fireground investigators have spent days searching the fireground while heavy cranes removed the debris, layer by layer.

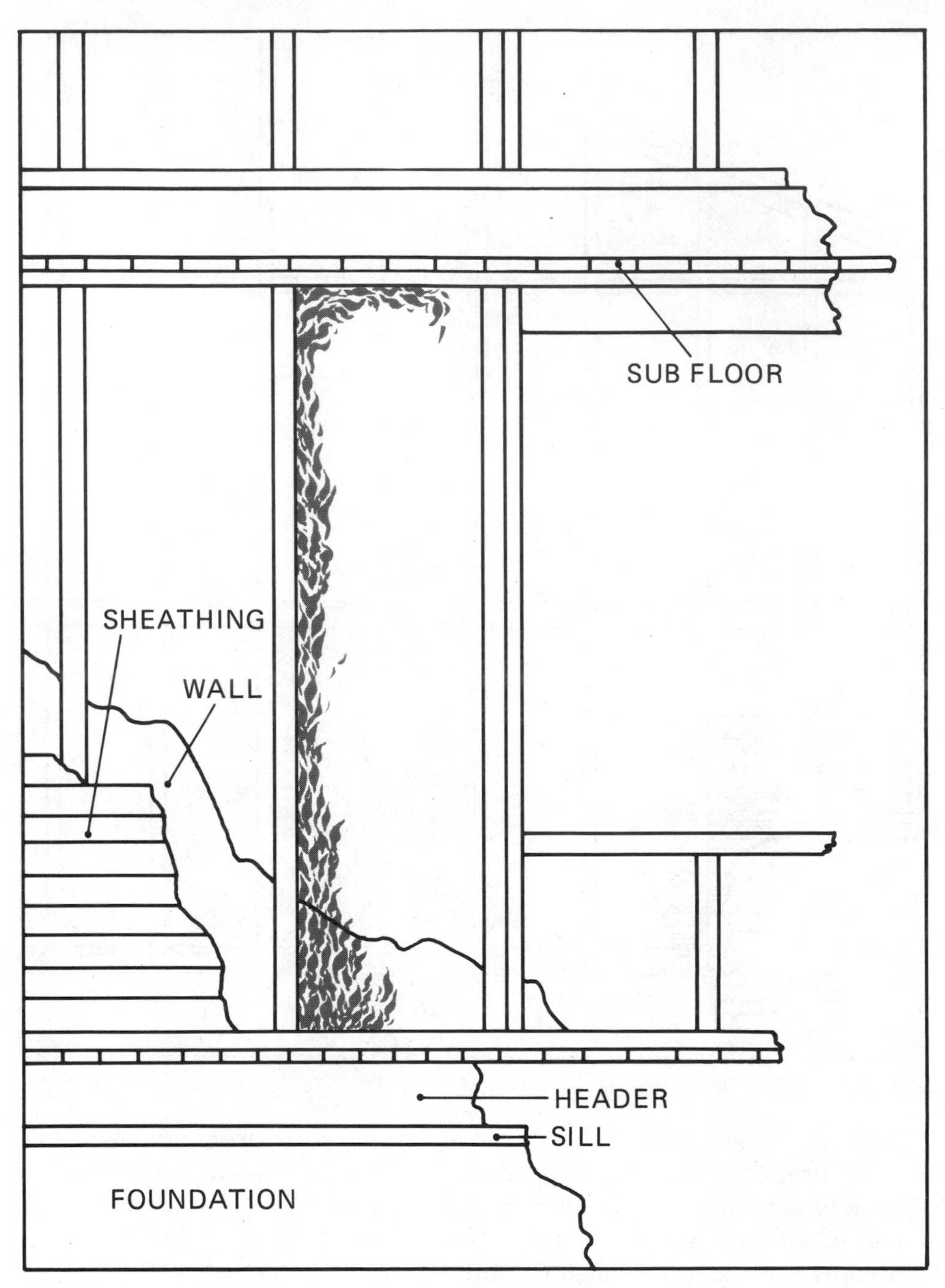

Figure 3-2. *Platform frame construction.*

Plank and Beam Construction. In this type of construction, the frame is made up of vertical and horizontal wooden beams. Beams may be formed into trusses to support sloping roofs. The beams are heavier than standard studs and are placed further apart than studs (four-foot and eight-foot

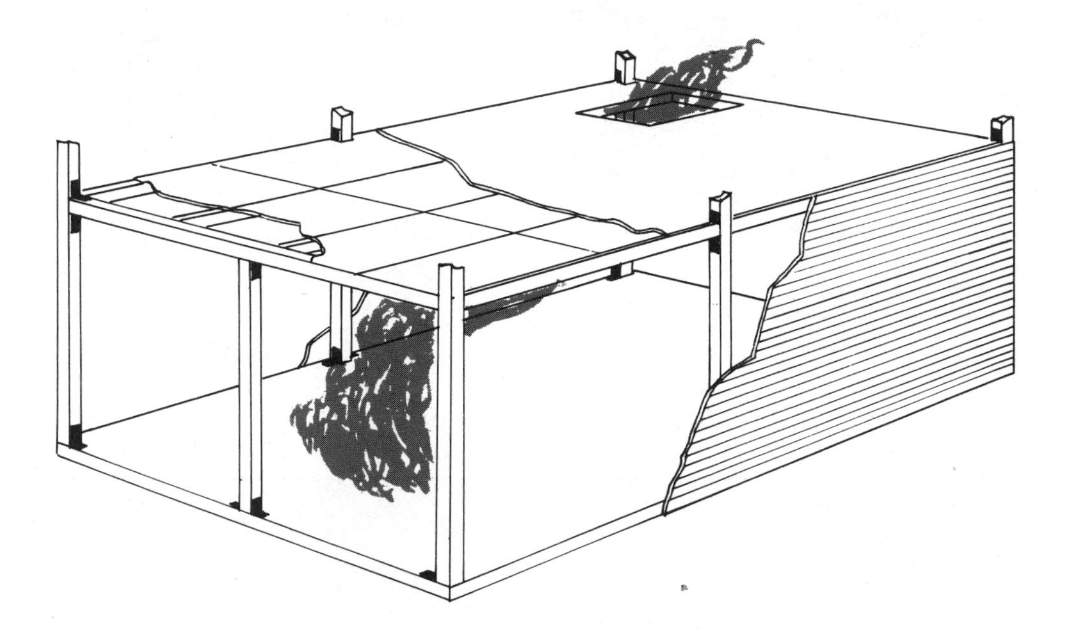

Figure 3-3. Post and frame construction.

spacings are not unusual). The walls are made up of thick wooden planks that span the spaces between beams (Figure 3-4).

The planks and beams are often left open to view from the inside of the structure. Then they may be protected from decay by highly flammable coatings of shellac, varnish, or (nowadays) polyurethane. Or the walls and ceilings may be covered with such materials as wood or plastic paneling, fabric, or sprayed acoustical foams consisting mainly of polyethylene. Both types of interior finish materials can promote the rapid spread of fire.

Ordinary Construction. In ordinary construction, the walls are of masonry and other structural elements are of wood (Figure 3-5). This construction method offers a variety of designs and low maintenance costs—a combination that is attractive to many businesses and commercial real estate developers. It also produces many void spaces between floors and ceilings and within interior walls. Fire can burn in these spaces for hours before it is discovered. Fire can also spread rapidly through these spaces, from room to room and from floor to floor.

Heavy Timber Construction. Heavy timber construction is characterized by massive structural members (Figure 3-6). Originally, this type of construction was used to support the heavy equipment in mill buildings, and it is sometimes called "mill construction." Fire can burn for long periods of

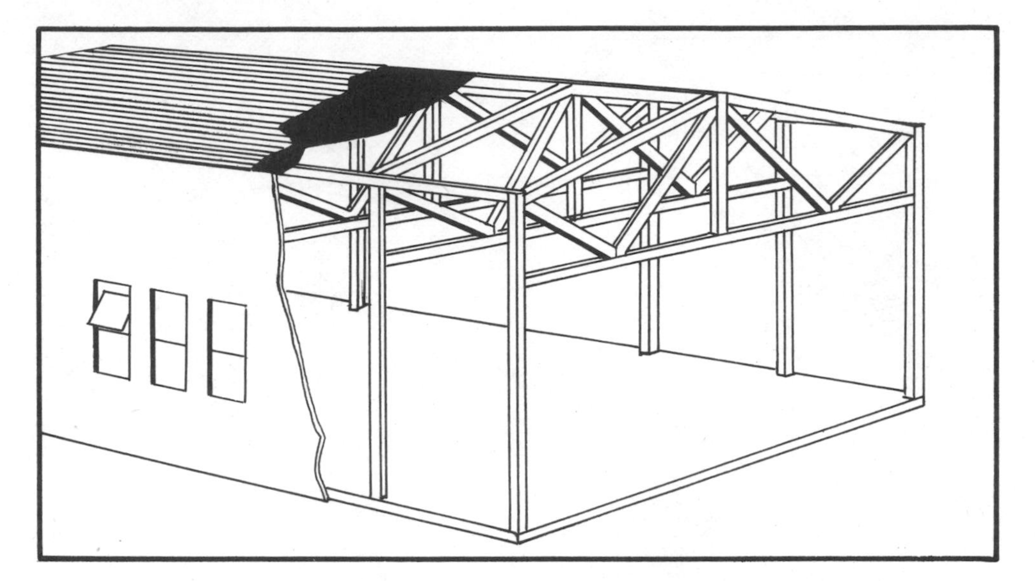

Figure 3-4. *Plank and beam construction.*

Figure 3-5. *Ordinary construction.*

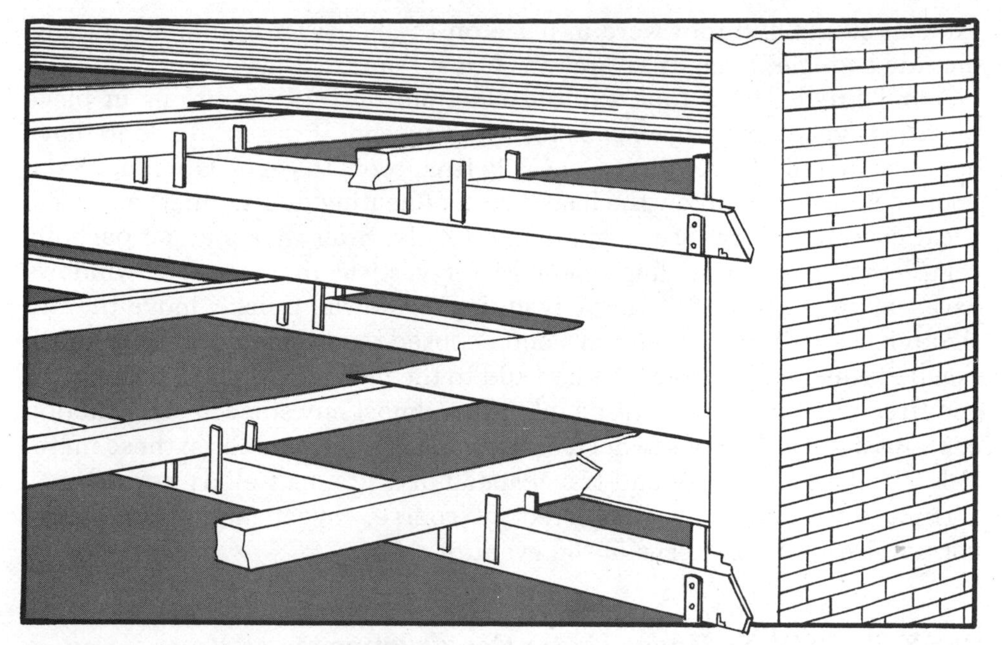

Figure 3-6. Heavy timber construction.

time in a building of heavy timber construction, without causing structural weakness or collapse. The fire will, however, move through openings in walls or floors and through all other available avenues of travel.

Combinations and Variations. Many buildings will exhibit a combination of construction types. This is especially so if a building has been enlarged or renovated. One part of the building may be of, say, balloon frame construction, and another, newer part of ordinary construction. Or a building may seem to be of post and frame construction from the outside, but the interior may contain several stories, each partitioned into a number of offices. Occasionally, an older building is modernized with a new exterior—perhaps a glass wall—or with a courtyard or atrium.

In each case, fire behavior will be influenced by the availability of fuels, channels for travel, and oxygen. The investigator's knowledge of fire spread in "pure" construction types can be used to reconstruct the path of the fire in a building of modified construction, and to determine whether that path was indeed a normal one.

Buildings Under Construction. Buildings in the process of construction present varied fire-travel situations. A steel and concrete building in the early stages of construction would probably not fuel a fire that started, say, in a pile of debris or construction materials. Without its exterior skin, the building would allow full venting of smoke and gases. However, a wood frame house could well become involved at this stage. If, in addition, the exterior siding and roof were in place on the house, it would, essentially, provide a single, large open area for fire spread.

A building with its roof, exterior skin, and interior partitions in place provides everything a fire needs for rapid spread. If the inside doors have not been installed, fire can travel easily from room to room. Openings from floor to floor, cut to allow the installation of heating ducts, pipes, and electrical conduits, allow fire to spread vertically. Stairways may be partially completed, again providing openings for vertical fire travel. If windows have not yet been installed, the resulting drafts will quickly move the fire through the building. Open walls and exposed studding may provide additional routes for fire travel. And, to add to the problem for both firefighters and fireground investigators, a building in almost any stage of construction contains a great deal of sawdust, scrap wood, and other debris; these materials affect the intensity and travel speed of a fire, as well as its path.

Once the building is completed, of course, investigators can expect normal fire-travel patterns in the event of a fire.

CASE HISTORY: House Under Construction

Firefighters responded to a structural fire in a residential area early on a cold March morning. The involved structure was a duplex apartment building in an early stage of construction. Exterior doors had not yet been hung, and the windows had only been nailed in place temporarily. These conditions allowed full draft and fast burning.

After extinguishment, a fireground investigation was begun. Investigators found two separate points of origin—one in the basement and one on the first floor. In the basement, the point of origin was a kerosene heater, tipped over onto a pile of marsh hay. On the first floor, the fire had originated in a pile of wood scraps, next to a pile of prefinished wall panels. Investigators could find no possible heat source at or near the first floor point of origin.

The construction foreman was called to the scene and questioned. He said that the kerosene heater had been used to dry the concrete walls in the basement; however, it had been moved from where he had placed it before leaving the job on the previous evening. The marsh hay was also out of place; it had been spread on the basement floor to prevent freezing.

Later questioning of the owner of the structure revealed that there had been a problem concerning the construction of rental property in a single family residential area. Neighbors had petitioned against the duplex; there had been a number of arguments involving the owner and neighbors, as well as petty damage to building materials.

The existence of two points of origin, the missing heat source, and the statements of the foreman and owner indicated arson. The investigators found as follows:

Points of origin: Basement, pile of marsh hay; first floor, pile of wood scraps
Heat source: Basement, kerosene heater; first floor, direct flame contact
Reason: Deliberate ignition
Category: Arson.

The area was declared a crime scene, and the arson squad was called in. The perpetrators were eventually found and convicted.

Interior Features and Space Configuration

The interior space configuration usually determines the burning rate. Fire travels rapidly in open, unobstructed spaces, provided there is something to burn. If the fire load is heavy and conducive to rapid burning, the fire will quickly engulf an open space, producing a massive fire. If the space offers an avenue for fire travel to an uninvolved space, the fire will do so. A crevice in a doorway, an air-conditioning duct, or an ill-fitting cover plate on an electric outlet will give fire the chance to advance from one area to another.

Small, divided spaces tend to slow the extension of fire. However, the fire will travel through any opening in a dividing wall or partition. Doors left open for convenience or blocked open will facilitate fire spread. Fire-resistant construction materials will help slow the spread of fire in a building that is subdivided into small spaces. In large, open spaces, fire will travel very rapidly, no matter what type of construction materials have been used.

Building features determine the predominant avenues of fire travel. In a single-story building, the fire travels horizontally until it vents itself; the only path to fuel and oxygen is a horizontal path. In a multistory building, the fire will travel upward if it started below the top floor. If the building contains many vertical shafts, the fire will travel these shafts to upper levels. If the building is tightly constructed of fire-resistant material, upward travel will be minimized. However, even in this case, the vertical movement of a large fire is rarely stopped completely. An air-conditioning duct or pipe shaft may provide a path for vertical extension. Open or heat-damaged windows may allow the fire to extend out the building at one level, and back in at a higher level. (This mode of fire travel is called *leapfrogging*.)

Many newer high-rise structures are built with a central utility core, containing elevator shafts, air-conditioning and heating ducts, piping, and electric wiring. Because all vertical shafts are confined to the core, upward fire travel is minimized. However, vertical fire spread is enhanced if the fire enters the core or stairwells located on outside walls.

Finishing Materials

The interior finishing materials are often the most hazardous combustibles in a building, in terms of their effect on fire travel. Walls, ceilings, and floors may be covered with such highly combustible materials as paint, varnish, lacquer, wallpaper (some with plastic coatings or bases), wood and plastic paneling, carpeting, foam rubber padding, and ceiling tiles. Although these materials burn at different rates, they all burn very rapidly at temperatures above 600°F (315°C).

On walls, ceilings, and floors, these materials present a large surface area to the fire and its hot combustion products. Radiant and convected heat from even a small fire will drive flammable vapors out of finishing materials, especially those used to coat the ceiling above the fire. A larger fire may produce flammable vapors across the walls, floors, and ceilings. When the temperature of these vapors reaches the ignition temperature, they will ignite all at once. The fire will flash over the finishing materials almost instantaneously. Thin materials, such as paint and wallpaper, will be consumed quickly; wood paneling and rugs will continue to supply fuel to the fire for longer periods, because they have more bulk.

Burning finishing materials extend the fire to other available fuels, such as combustible contents. Finishing materials may also provide a "bridge" between the original fire and such means of fire travel as stairwells, ducts, floor and ceiling openings, and soffits.

Renovations. Most renovations or modernizations seem to increase the probability of an intense, rapidly spreading fire. The partitioning of previously large, open areas into smaller offices will, of course, restrict fire travel. However, a favorite renovation for commercial buildings seems to be to drop the ceiling and install wood paneling on plastered walls. The wood paneling increases the fire load and the burning rate. The dropped ceiling helps combustion products to accumulate faster (and get hotter), and it often becomes involved with fire. The lowered ceiling also creates a void space, between the old and new ceilings, that promotes the horizontal spread of fire.

Since renovations and modernizations are very much a matter of individual taste, each is more or less unique. The fireground investigator must, in each situation, determine how the renovation affected (or was made to affect) fire behavior.

BUILDING CONTENTS

In very general terms, bulky fuels (such as thick wooden furniture, bedding, baled goods, books, and stacks of materials like corrugated cardboard or plywood) give off flammable vapors only slowly. These bulky materials therefore burn slowly, producing deep but wide V-shaped burn patterns on themselves and on nearby walls. Less bulky or more flimsy materials (such as curtains, rugs, and individual sheets of cardboard or plywood) give off flammable vapors more quickly. They burn rapidly and, if not completely consumed, show shallow and narrow V-shaped burn patterns.

Flashover may occur and result in a hot rapid spread of fire. Flashover is when all combustibles in the area reach ignition temperatures at the same time and ignite. (Burn patterns are discussed in detail in Chapter 7.) Flammable liquids can, however, transform slow-burning contents into fast-burning fuels, producing abnormal burn patterns.

Pulverized materials (dusts) burn very rapidly, often with explosive force as noted in Chapter 2. Plastics and other synthetic materials may add to the complexity of analyzing fire behavior and burning rates; many of these materials contain oxygen that is liberated during combustion.

Normal Contents

The normal contents of structures vary with the type of occupancy and, in commercial and industrial buildings, often with the season. The type of contents is usually predictable with some degree of accuracy.

Dwellings. Dwellings contain furniture and other decorative items, clothing, and items related to maintenance and recreational activities. One home may contain an abundance of rugs and overstuffed furniture, while another is almost all glass, chrome, and wood-paneled walls. Basically, though, residential fireloads do not vary greatly—especially in dwellings of nearly equal size.

The placement of residential contents is also fairly standard, with some variation due to different tastes in decorating and differences in floor plans. Thus, fire behavior and the resulting burn pattern in an involved residence should approximate some normal pattern. Deviations would depend on the type of construction, firefighting operations, and the point of origin of the fire.

Commercial Buildings. Commercial and industrial installations vary widely in fireload. However, a haberdashery would be expected to contain men's clothing, and not sawdust. A pizza parlor should contain tables, chairs, pizza-making equipment and pizza ingredients, but not a dozen cans of kerosene. And so on. In each case, the normal contents, along with other known factors, should result in predictable fire behavior. When they don't, there is cause for suspicion.

Warehouses. Almost any type of material may be stored in a warehouse, and each type will produce different fire behavior. Moreover, the path and intensity of a fire in a fully stocked warehouse will be quite different from those in a mostly empty warehouse, even if exactly the same type of material is stored in both—and even if the two warehouses are exactly alike in construction. However, knowing the contents and other influencing factors, the investigator should be able to tell whether the evidence left by a fire is normal or abnormal.

Factories. Manufacturing installations contain finished goods, partially completed products, raw materials, processing equipment, and maintenance equipment and supplies. Many of these materials and supplies are dangerous. For example, a furniture factory might contain wood, glues, lacquers, foam rubber, bolts of cloth, and a hazardous by-product, wood dust. All these materials are extremely flammable, but they are also normal to furniture production (Figure 3-7).

Figure 3-7. *High-hazard contents result in a hot, rapid-moving fire.*

Similarly, even the most diversified manufacturing plant contains the normal complement of equipment, goods in process, materials, and supplies. And each of these should be found in the plant in normal amounts. If such a plant becomes involved with fire, the resulting fire patterns should be normal for that installation.

Hazardous chemicals and incompatible chemicals (which may ignite simply by coming in contact with each other) are being used in increasing amounts in industry. New chemicals and new uses for chemicals are introduced almost daily. Often, the producers are unable to supply guidelines for safety in using these chemicals. An investigator should seek help immediately (request an advanced investigation) when a fire involves such hazardous materials.

Seasonal Variation. The fireloads of commercial and manufacturing installations may also vary with the season of the year. What are normal contents in one season may not be normal in another. For example, retail stores change inventories with the seasonal demands of their customers. Toys stocked for sale at Christmas, being largely made of plastic, would produce fire behavior different from that of, say, lawn and garden tools stocked for sale in the spring.

Warehouses and factories may be filled to overflowing in the weeks before a particular selling season, and empty once the season starts. And the type of contents stored or manufactured may vary with the selling season. In each case, it is the investigator's responsibility to determine what was housed in the structure at the time of the fire, and whether the course of the fire was consistent with those contents.

Normal Versus Abnormal Contents

It seems that almost anything can be part of the normal contents of a structure. Yet the fireground investigator must seek and find any abnormal items or substances at the fire scene. What should be considered abnormal, if almost anything can be normal? This question can be resolved to a degree by

1. Determining what is normal for the particular occupancy
2. Looking for contents that are out of context with the occupancy
3. Determining why these seemingly abnormal contents are present.

It is of help to consider the circumstances and the quantity in which the abnormal contents are found. For example, we have already discussed the normal contents of the usual residence. A supply of gunpowder would be out of context there. However, the presence of the gunpowder would be explained if the occupant was a hunter who made up his own shells. A small amount of gunpowder that became involved in a basement fire would then be considered normal. But a trail of gunpowder leading through the residence from the fire's point of origin would be far from normal.

This type of reasoning can be applied to commercial and industrial installations, but much more knowledge is needed. The fire investigator cannot be expected to know what goes on in every factory in a sizeable fire district. However, preplanning and fire-prevention inspection reports contain much useful information. In addition, arrangements can be made to visit the sites of unusual manufacturing processes, to gain first-hand knowledge concerning their normal contents.

FLAMMABLE AND COMBUSTIBLE LIQUIDS

Flammable and combustible liquids are used by arsonists as accelerants, to increase the intensity of a fire and the speed at which it spreads. They may also be part of the normal contents of a structure. It is up to the investigator to determine whether a flammable and combustible liquid was involved in a fire, whether it was responsible for rapid spread of the fire, and whether the involvement was accidental or intentional.

Flammable or combustible liquid fuels are quickly vaporized and consumed by fire; however, they leave burn patterns and other evidence that is fairly easy to detect. (This is true even when a professional arsonist uses accelerants to move the fire rapidly along a path that is consistent with normal fire travel.) In this section, we discuss the detection of flammable and combustible liquids in some detail.

An important point to remember is that hydrocarbons do not ignite spontaneously. They are quite dangerous when left in open containers, because they vaporize readily at normal temperatures. But even then, they must be ignited by a source of heat.

Smoke

Normally, a fire in a residential structure produces a gray smoke at first. As the fire progresses, the smoke becomes darker. When the fire reaches the roofing materials, it gives off a heavy black smoke that is usually associated with petroleum products, or hydrocarbons. This follows from the fact that asphalt shingles and roofing tar have a petroleum base.

Most common flammable and combustible liquids, including lighter fluid, gasoline, and kerosene, are also petroleum-based products. They produce the same type of dark, heavy smoke (and red flames) when they burn. They also produce a hot, fast-burning fire that may leave characteristic cracks on glass windowpanes (see Chapter 7).

The fireground investigator should determine—from firefighters, witnesses, or, if possible, personal observation—the color of the smoke during the early stages of a fire. Black smoke, perhaps combined with evidence of intense heat, would indicate that hydrocarbons were involved. These hydrocarbons need not be in the form of flammable or combustible liquids; for example, the plastic upholstery on a chair and the foam rubber in a mattress are also petroleum-based products. However, the investigator can find out what produced the black smoke only by carefully examining the involved area.

Odor

Accelerants have strong, characteristic odors. In many cases, the odor of a flammable and combustible liquid will linger near the point where it was consumed. The odor will remain for only a limited time after extinguishment. However, it may be detected by investigators if the examination of the fire building is started immediately following extinguishment. Or, the odor of a flammable or combustible liquid may have been noticed by advancing firefighters, who should be asked about it directly.

Accelerants can be distinguished, one from the other, by smell. Lighter fluid, for example, leaves a sweet, lingering odor for a short time. Gasoline and kerosene can usually be distinguished from other accelerants and from each other by their odors. Fire investigators should become familiar with the odors of the various flammable and combustible liquids, as an aid in detecting and identifying them. An odor is not evidence of the presence of flammable and combustible liquids; however, it is of help to the investigator.

Hydrocarbon Detector

A more accurate method of finding flammable-combustible liquids is by using a hydrocarbon detector. This instrument can detect hydrocarbon vapors in very small quantities. Although there are several types, they are all used in about the same way. A hollow probe is moved about the area in question; the probe picks up any hydrocarbon vapors in the area and registers their concentration on a gauge in parts per million (PPM) in air.

Hydrocarbon detectors were developed mainly for use in emergency situations. They can, for instance, be used to determine whether a leak of hydrocarbons has resulted in an explosive vapor-air mixture. But they cannot distinguish among the different hydrocarbons. Full instructions are provided when a detector is purchased.

Burn Patterns

When flammable or combustible liquids burn, they leave a distinctive pattern, unlike the patterns of other combustible products normally found in a structure (Figure 3-8). There are several differences.

First, the ignition of flammable-combustible liquids in a room results in *even* smoke and heat patterns on the surrounding walls. Fire in, say, a pile of combustibles produces heat and smoke stains that are heavier on one wall than on the others.

Second, when a flammable-combustible liquid is involved at the outset of a fire, a definite, single point of origin is not produced. The flammable-combustible liquid will be poured or spilled over an area—say on a sofa or a floor. Upon ignition, flames will appear above the entire area covered by

Figure 3-8. *An indication of an accelerant is the intermixing of shallow and deep erratic ground lines.*

the accelerant. This results in an *area of origin,* with the entire area showing damage of equal magnitude. On the other hand, fire in ordinary combustibles begins at one point—the point of contact between the combustibles and the heat source. That point, the point of origin, shows the most fire damage and is usually the lowest point at which damage occurs.

A third difference is in the burn marks themselves, especially on floors. We have noted that ordinary combustibles leave V-shaped burn marks. On a tightly constructed floor, flammable or combustible liquids tend to form pools, or puddles, when they are poured. Upon ignition, the flammable vapors burn, but the liquid accelerant does not burn. Instead, the liquid insulates the floor surface directly below it. Only the flooring around the edges of the pool are burned (Figure 3-9). Then, as the liquid continues to vaporize, the pool becomes smaller; more of the flooring around the edges of the pool is exposed to the flames and is burned. This process continues until all the accelerant has been vaporized. Then, what is left is a residue and a char or burn pattern in the shape of the original pool. The pattern is usually oval and irregular in outline. Residue is discussed in the next section.

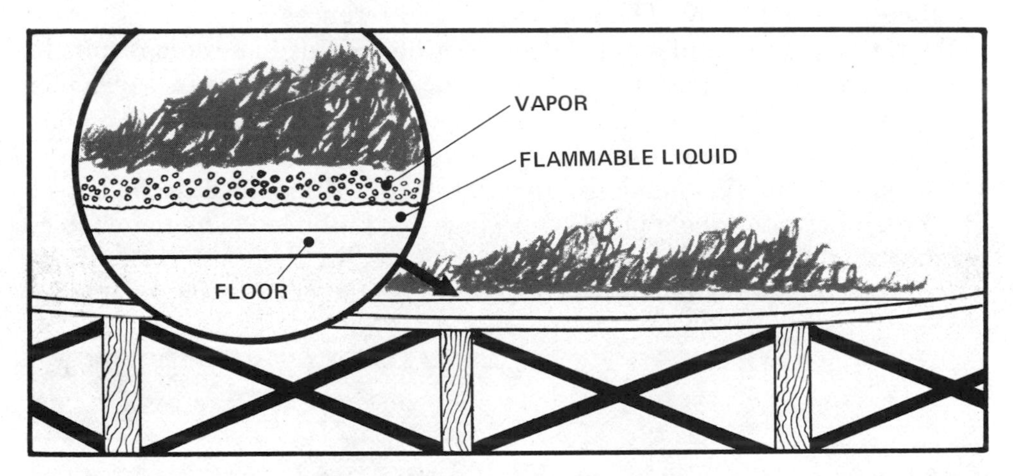

Figure 3-9. *The puddle of flammable liquid insulated the floor from heat.*

A flammable or combustible liquid rarely causes the fire to burn down through tight flooring. Instead, flames from the burning liquid contact nearby combustibles, igniting them; the fire then travels in the usual upward direction. However, if the flooring is loose, perhaps because there are cracks between the floorboards, liquid accelerant or heavy vapors can drop down. Ignition can then occur in the floor joists, creating an upward burn. The burn pattern would indicate burning upward from below the floor. Thus, the under side of a floor should always be checked for burn patterns. If the floor has not burned through from the top, but there are burns under the flooring, a flammable or combustible liquid may have carried the fire down below the floor.

On a porous but nonflammable material such as concrete, an accelerant will soak in, eventually vaporize and burn, but will leave a detectable residue. Obviously, the concrete will not be damaged.

A fourth difference between the burn patterns of flammable-combustible liquids and those of ordinary combustibles stems from the liquids' ability to carry fire downward. Liquids tend to seek the lowest level; they will run downward if poured on, say, a sloping floor or a stairway. If a flammable-combustible liquid is ignited at the high point, the flames will flash downward along the path of the liquid. In addition, where less liquid has collected, the accelerant vaporizes quickly. This leaves a series of pools of liquid that burn for a longer time, and that produce a line of oval char patterns. These patterns can be recognized as resulting from the flow of flammable or combustible liquids.

A flammable or combustible liquid that has been poured or spilled on a floor may flow under a nearby door. When ignited, the liquid will char the bottom edge of the door (Figure 3-10). Since only a flammable or combustible liquid will leave this type of burn mark, it is a good idea to check doors in the involved area.

Flammable or combustible liquids also leave distinctive burn marks on ordinary combustibles. The liquid is absorbed into the material, causing deep burning. The burn pattern is irregular in shape; residue may remain even if the material is consumed (Figure 3-11).

Residues

Flammable or combustible liquids usually leave physical evidence of their presence, even when they seem to have been consumed completely. That evidence is a residue, left by the liquid that has soaked into any porous surface. (Flammable or combustible liquids do not leave a residue on surfaces that are absolutely nonporous. However, such surfaces are very rare, and a porous surface can usually be found nearby.)

The residue can be found by carefully sifting through and examining the debris at the suspected area of origin. If possible, debris containing the residue should be recovered and preserved as evidence. A simple but effective way to detect the residue is as follows: Place a piece of debris from the area of origin in a container of clean water. If flammable or combustible liquid has soaked into the debris, a film will appear on the surface of the water. If the film appears, the specimen should be sent to a laboratory for analysis. (See Chapter 8 for details regarding the preservation of evidence and the submission of evidence to testing labs.)

Puddles of water in the area of origin should also be examined. A film on the surface of the water is most likely the residue of an accelerant. Again, a sample (of the water and film) should be obtained in a clear jar and sent to a laboratory for analysis.

Figure 3-10. *Burn patterns from spilled flammable liquids: notice bottom edge of door and vertical burn patterns on both sides.*

CASE HISTORY: Flammable Liquid

Fire companies responded to an apartment fire on a hot July evening. On arrival, they observed very dark smoke and reddish flames issuing from

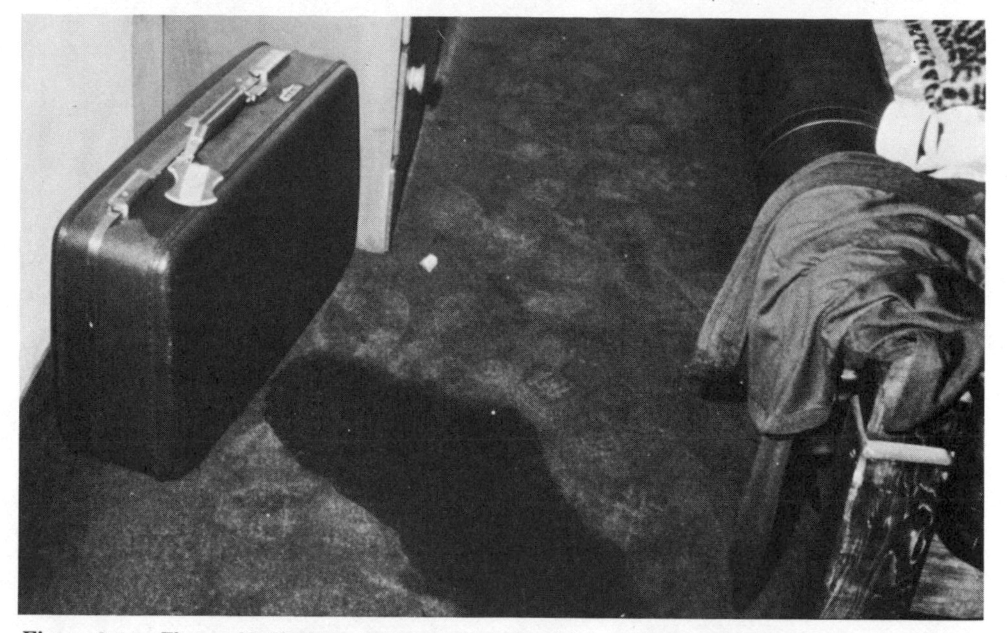

Figure 3-11. *Flammable liquid soaks into porous materials, such as the rug.*

the apartment's kitchen windows. The firefighters were able to confine the fire to the kitchen area and extinguish it quickly.

The fire patterns led investigators to a storage closet off the kitchen. The closet door was in the closed position, and only the bottom third remained. The top two-thirds of the door had been consumed, allowing the fire to flash across the kitchen ceiling. The bottom edge of the door was deeply charred, indicating that a flammable liquid must have run under the door. Immediately in front of the door was an irregular oval flammable-liquid burn pattern. The debris in the closet included a ruptured can that had contained a flammable liquid. The occupants were asked what items were stored in the closet; among them were a can of charcoal lighter fluid and a bag of charcoal.

After further questioning of the occupants, investigators were able to reconstruct the events leading to the fire: The male occupant had taken the bag of charcoal and lighter fluid outside to start his grill for supper. He filled the grill with charcoal, used the fluid, and lit the fire. After a while, he stirred up the charcoals in the grill, picked up the charcoal bag and can of fluid, and carried them inside, and placed them in the closet. During the stirring of the grill, a hot piece of charcoal had fallen into the open bag. In the closet, the bag smoldered and burst into flame, igniting surrounding combustibles. The heat ruptured the can of lighter fluid, accelerating the fire and accounting for the evidence of flammable-liquid involvement. The fire investigation report listed the following:

Point of origin: Corner of kitchen storage closet

Heat source: Glowing charcoal
Reason: Hot coal dropped into charcoal bag
Category: Accidental.

FIREGROUND OPERATIONS

Standard firefighting tactics, properly applied, will not cause unusual fire behavior. That is, the smoke and burn patterns produced during attack, venting, and extinguishment will be those normally produced. They will be expected by investigators and attributed to the firefighting operations.

However, firefighting operations are not performed in the calmest of environments, and errors do occur. Improper firefighting tactics result in erratic or abnormal fire travel, which can leave confusing smoke and burn patterns. The investigator "reading" these patterns can be led to false conclusions unless he knows what caused them.

For this reason, it is important that the fireground investigator know what procedures were used in attacking, venting, and extinguishing the fire. It is best if the investigator is at the scene to see the attack, but this is not always possible. Next best is to question firefighters about the tactics they used—after the fire has been extinguished. In addition, the investigator can consult fire reports. However, these reports are often too sketchy to be of help. As part of any anti-arson effort, therefore, fire reports should be as complete as possible. They should include the tactics employed in fighting the fire and the points at which such tactics were employed (Figure 3-12).

Attack

A common mistake is sometimes made by hose men, outside the fire building, who find fire coming out a window. Their first reaction (an improper one) is to attack by directing their streams at the window from outside. This action pushes the fire back into the building and into uninvolved areas (Figure 3-13). The proper tactic is to attack the fire from inside the building, drive it toward the window, and extinguish it.

A similar error is an attack from or at the roof after a fire has vented itself; again the fire is driven down into the building and toward uninvolved areas. Pushing the fire along a narrow hallway has the same effect. In all these cases, confusing burn patterns will make fireground investigation extremely difficult.

Ventilation

Perhaps the worst venting error (for both firefighters and investigators) is to make entry into an oxygen-starved atmosphere without first ventilating the building properly. The resulting backdraft explosion could cause severe injury, cause further damage to the fire structure, and accelerate burning

COMPANY OFFICERS REPORT
GENERAL ALARM

ALARM NUMBER 893 **COMPANY** 1 **DATE** 10/7 **19** 80
LOCATION OF ALARM 2130 Olive Tree Drive
RESPONDED FROM Central Fire Station
DELAYED: YES___ **NO** X **REASON** ---- **TIME** 1:43 A.M.
EQUIPMENT USED:

EXTINGUISHERS: CO2___ **ABC**___ **H20**___ **HALON**___
PUMP YES X **NO**___ **TIME:**___ **HRS** 5 **MIN**
HOSE:

SIZE HOSE	FOOTAGE WET	DRY	FROM	TO	PUMP PRESSURE	ESTIMATED GALLONS
1″						
1½″	200′		Pump	Fire	150	180
2½″						
3″						

OPERATIONS

SKETCH

N
W
E
BR
BR
K
BR
BA
LR
VINE ST
ENG 1
1-1/2″ LINE

STORY
Upon arrival we found smoke heavy in the garage area. The occupants were outside and informed me that the fire was in the garage. The driver set the pump, I notified the shift commander of the situation and along with my firefighter took the pre-connect 1½″ into the front door. We advanced to the kitchen door that leads to the garage. The fire was in the south-east corner at the front of the auto. We extinguished the fire and opened the large overhead door. Truck 9 and Engine 7 personnel helped push the auto out of the garage. Following extinguishment, we picked up our equipment and were sent back to quarters.

RESPONDING PERSONNEL: Captain Herb Polsin
Driver Ned Nactman
Firefighter Tim Carter

OFFICER IN CHARGE: Captain H. Polzin

Figure 3-12. Company Officer's Report: General Alarm.

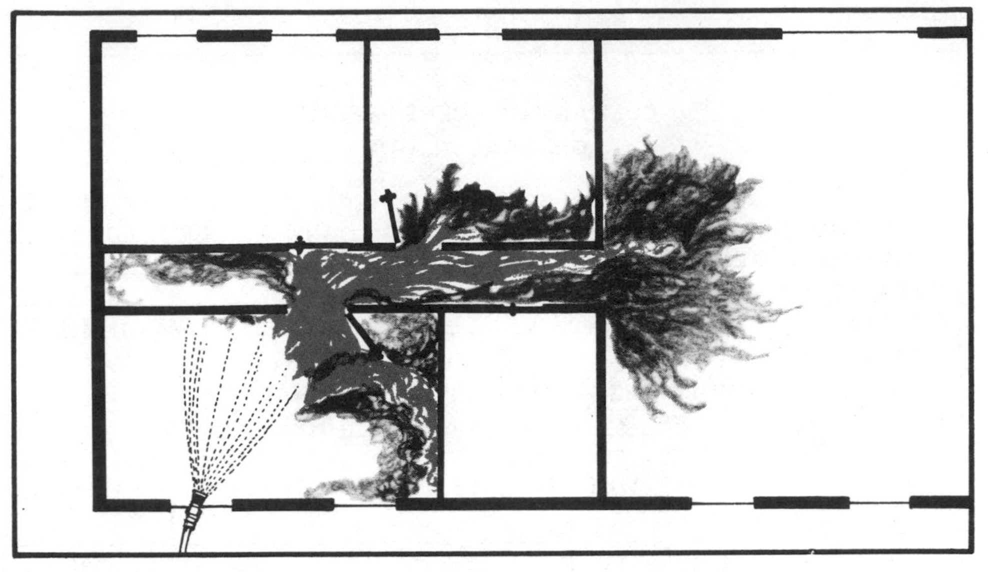

Figure 3-13. Improper tactics.

and the extension of the fire. The excessive damage and burning would add to the difficulty of fireground investigation and, again, could lead to false conclusions.

Ventilation influences the path taken by a fire. When an opening is made, air is drawn toward the opening from all parts of the building. This forms a draft that pulls the fire along the shortest path to the opening (Figure 3-14a). The fire is accelerated as it moves toward the opening, but it is kept from spreading outside the path.

Ventilation openings must, therefore, be properly placed with regard to the fire situation. For example, a roof opening that is not directly over the fire will draw the fire laterally across ceilings before it moves vertically to the opening (Figure 3-14b). Windows opened at random may do nothing at all, or they may push the fire into uninvolved areas. When cross ventilation is used, the leeward side of the building must be opened first; then the windward side may be opened. If the windward side is opened first, hot combustion products are pushed along new avenues of travel in the building; they have nowhere else to go.

Again, these ventilation errors do occur. The important point is that the investigator needs to be aware of fireground mistakes. Firefighters can be of help because they have this information, and investigators can help themselves by asking.

EXPLOSIONS

Explosions were discussed in Chapter 2 and will not be discussed in detail here. They can, however, scatter or destroy evidence, making investigation very difficult. The investigator should first attempt to determine where the explosion occurred. Then he should determine what exploded, and why.

In the basement of a residence, for example, a defective boiler or a gas leak might be a logical cause of an explosion. A gas leak might also cause an explosion on an upper floor. Backdraft is always a possibility, but it usually does not have the force to knock a building down. In a manufacturing plant, causes could range from explosive dusts to flammable or combustible liquids to dangerous processes such as spray painting. If there is no possibility of an accidental cause, the investigator should look for evidence of a bomb explosion.

Finally, we repeat that in the case of an explosion combined with fire, the investigator should try to determine which came first. This information is very often of help in determining the causes of both.

LOW BURN AND POINT OF ORIGIN

In this chapter and the previous chapter, we have noted that:

- A fire starts when a source of heat raises the temperature of a fuel to or beyond its ignition temperature.
- If the fuel is ordinary combustibles, the fire begins at a point; if the fuel is a flammable or combustible liquid, the fire is initiated over an area.

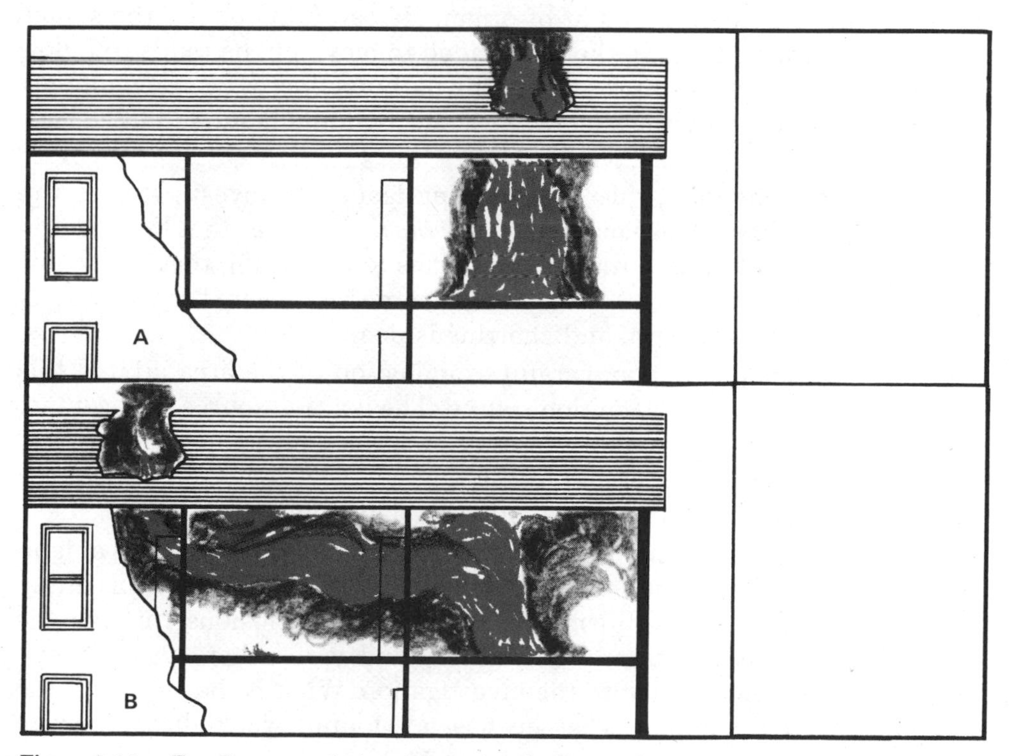

Figure 3-14a. *Top: Proper ventilation.* ***Figure 3-14b.*** *Bottom: Improper ventilation.*

- The fire starts small and grows larger if fuel and oxygen are available to it.
- Fire spreads upward where that is possible; laterally where there are no vertical pathways; and downward rarely, unless it is following a flammable or combustible liquid path or dropping embers.
- As the fire grows, it leaves a V-shaped burn and/or smoke pattern.
- Fire behavior and fire spread are influenced by building construction, fire load, the presence of flammable or combustible liquids, firefighting operations, and explosions.

These are general statements, and they represent quite a bit of detail. The details come into play at the fireground, where the investigator must deal with specific construction features, fuels, avenues of travel, and so forth, to determine the specific point of origin, heat source, and reason for the fire, and then categorize it correctly.

But where does the investigator begin, and how? He begins by locating the point or area of origin—the exact place where the fire began. He does this by moving along the path taken by the fire, but in the *opposite* direction. The investigator doesn't know where the fire started, but he knows where it was extinguished. He also knows that the fire probably burned longest and did the most damage at the point of origin. He therefore starts where the fire was extinguished, or where it did the least damage, and works back toward the unknown point of origin. He carefully follows the smoke and burn patterns; because the fire tended to move up, he tends to follow the patterns downward. As he traces the path backward, he takes into account the influence that available fuels and building features might have had on the fire.

In the ideal situation, this backtracking leads the investigator to the lowest point of fire damage, called the *low burn.* He notes that beyond this low burn, the smoke and burn pattern begins to rise again; that is, the low burn is at the point of the V-shaped pattern. Moreover, fire damage is substantial at the low burn, and charring is deep.

Still in the ideal situation, careful examination of the area around this low burn yields a possible ignition source. The ignition source is in contact with the remains of flammable materials at the low burn, or obviously was in contact with these materials. And examination of the remainder of the fire building does not turn up any other likely low burns.

The investigator is then fairly sure that the low burn is the point of ignition, and that he has found the source of ignition. As he continues his investigation, he will try to confirm these tentative conclusions through the questioning of occupants and firefighters and with other available evidence. But he now is well into the investigation. What he has found at the low burn (and sometimes what he has not found) will indicate how to proceed.

Relation Among Low Burn, Ignition Source, and Point of Origin

Very few investigative situations are ideal. Yet experienced fireground investigators can find the low burn almost every time, using the method outlined above. They can distinguish between the *true* low burn and *false* low burns created by dropping embers. And they can tell when two or more low burns were produced by an arsonist. Chapter 7 provides many more details about the procedure.

Here, we wish to stress one point: A low burn cannot be the point (or area) of origin of a fire unless there was contact between a combustible material and a heat source at that point. (In the case of ignition by radiated heat, sufficient heat must have been radiated to the combustible material.) The investigator must be able to logically connect the heat source and the fuel at the point of origin. Heat sources are discussed in detail in the next chapter. The remainder of this chapter is concerned with the low burn and point of origin.

Low Burn

By carefully tracing burn patterns back along the path of the fire, the investigator may discover one or more low burns. Each low burn may or may not be directly associated with an ignition source.

Single Low Burn. The simplest situation is that in which a complete examination yields one low burn along with a directly associated ignition source. The investigator is then justified in assuming that he has found the point of origin and heat source.

In some cases, the single low burn is found quickly, but the heat source is not evident. For example, suppose a wall is partly burned away at the low burn. The heat source may have been faulty wiring within the wall or something outside the wall, in the room itself. As another example, there may be no obvious heat source within a reasonable distance of the low burn. This may indicate that the fire was the work of an arsonist, or it may mean that the heat source is buried under debris, was consumed by the fire, or was shoveled out of the building during overhaul and cleanup. The debris near the low burn, and any debris that was moved away from it, must then be searched for a possible heat source.

An arsonist must combine a source of heat with fuel to start a fire. He can start the fire at an available heat source, or supply the heat source. If he chooses an existing heat source, he must move the fuel to that heat source. The fuel can then be recognized as being out of place. If the arsonist supplies the ignition source, the low burn will not be related to any of the fixed heat sources in the fire structure. This should alert the investigator to the possibility of arson.

Sometimes an arsonist will try to disguise the actual heat source. He may, for example, place crumbled-up paper on the floor beneath an electric outlet, and set it afire with a match. This, he hopes, will shift attention to the electrical system as the ignition source. However, the low burn should indicate to the investigator that the fire started below the outlet.

Two or More Low Burns. Some fires may produce evidence of several low burns. This is especially possible if the fire is a large one or if the fire structure is of balloon frame construction. The presence of two or more low burns may also indicate that an arsonist set the fire.

A fire that begins at one point of origin has one *true* low burn. This is the low burn that is associated with the point of origin and the heat source. As the fire progresses, it generally moves upward and out from the point of origin. As the fire intensifies, strong convection currents may lift and transport embers. These embers may drop down onto combustibles, igniting them and producing other, *false* low burns. To determine the point of origin and heat source, the investigator must be able to distinguish the true low burn from the false low burns. One way to do this is to "read" the smoke and burn marks and consider them in relation to influences such as building construction type. Often, this will reveal that embers could have carried fire from one low burn to another, but that fire travel in the opposite direction would be impossible. Since fire travels predictably, it should be obvious which low burn is false.

The smoke stains between two low burns may also be of help in determining which is the true low burn. Smoke stains usually slant upward and away from the point of origin—and thus *toward* the false low burn.

Generally, the lowest burn is the original or true low burn. Two low burns in the same general area may appear to be at the same level. Then the true low burn can be found by checking the depth of char at both low burns. A nail, penknife, or icepick is pushed into each char. The low burn that allows the deeper penetration has the deeper char and is the true low burn (Figure 3-15).

The true low burn must be directly associated with a heat source. If one of several low burns can be connected with a heat source but the others cannot, then there is no question which is the true low burn. For example, suppose a dishcloth is placed on a kitchen counter top near an active burner on a gas range. The flame from the burner ignites the dishcloth; flames travel upward, igniting a roll of paper towels, paint on the wall, and cabinets over the range. A burning piece of the dishcloth drops to the floor, where it ignites the lower cabinets and starts an upward extension (Figure 3-16). There are now two low burns—one at the original position of the dishcloth and one below it, near the floor. However, only the upper low burn is directly associated with a heat source; it must be the true low burn and the point of origin. The low burn near the floor, without a heat source, is a false low burn.

As another example, consider a situation in which electrical wires, behind

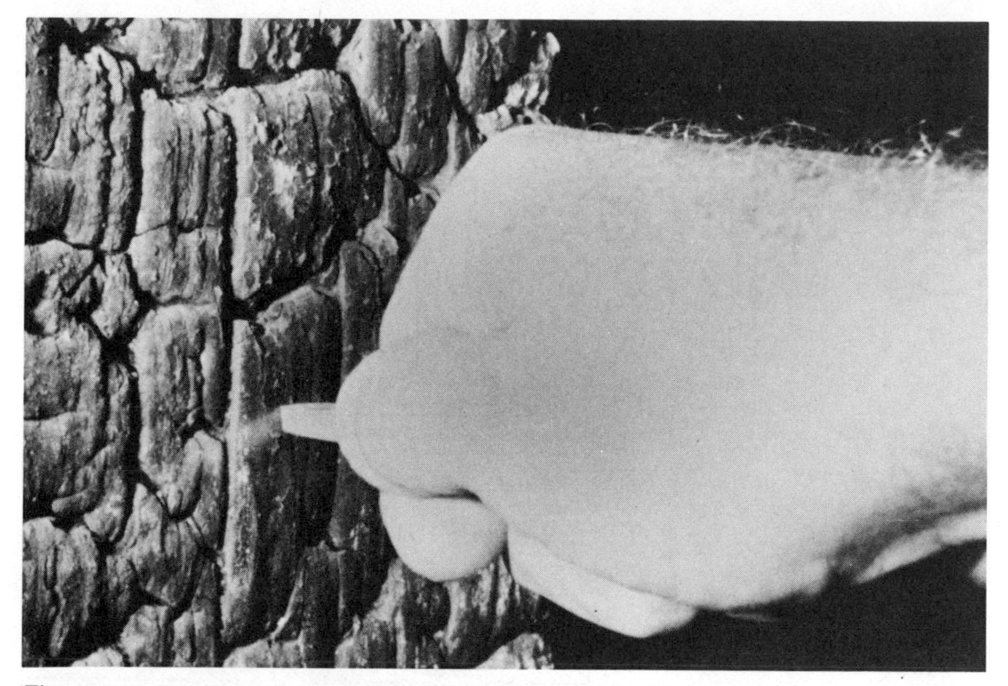

Figure 3-15. *The deeper the penetration, the longer the burn time.*

a wall outlet, overheat and start a fire. Smoke and flame issue from within the wall at the outlet cover plate. Because fire rarely travels downward, the point of the V-shaped burn pattern is at the cover plate (Figure 3-17). To confirm the heat source, the investigator must remove the cover plate and examine the wiring.

As the fire burns, embers may fall to the floor from the area above the outlet. They might ignite combustibles at the floor, which would then burn upward. The situation would be that shown in Figure 3-18, with the point of the V now at floor level. Inspection of the floor area would, however, show that there is no heat source at that level; inspection of the wiring would confirm the overheating there. The outlet would correctly be determined to be the point of origin.

It is, of course, possible to find two or more true low burns. This would indicate that separate fires had started in the same structure—usually as a result of arson.

False Indications of Low Burn. A visible V pattern may not always indicate the true low burn. A V pattern may seem to terminate at a wall shelf or above a piece of furniture, but may actually continue below it. The investigator must check to see what might be below the apparent point of the V.

For example, a fire caused by faulty wiring might start in the space between a ceiling and a floor, extend up through a pipe chase, and then involve a nearby wall (Figure 3-19). The visible V pattern would terminate at the floor, where the flames broke through the pipe chase. However, the true

Figure 3-16. Dish cloth is point of origin (true low burn) while floor pattern is false low burn.

low burn would be below the floor. Firefighters would probably have pulled the ceiling below to effect extinguishment, and the floor would be in place. The investigator would have to examine the area from below, to find the low burn.

Occasionally, a false low burn will seem to be "truer" than the true low burn. This is most likely to happen when the false low burn results from the involvement of more highly combustible materials than the true low

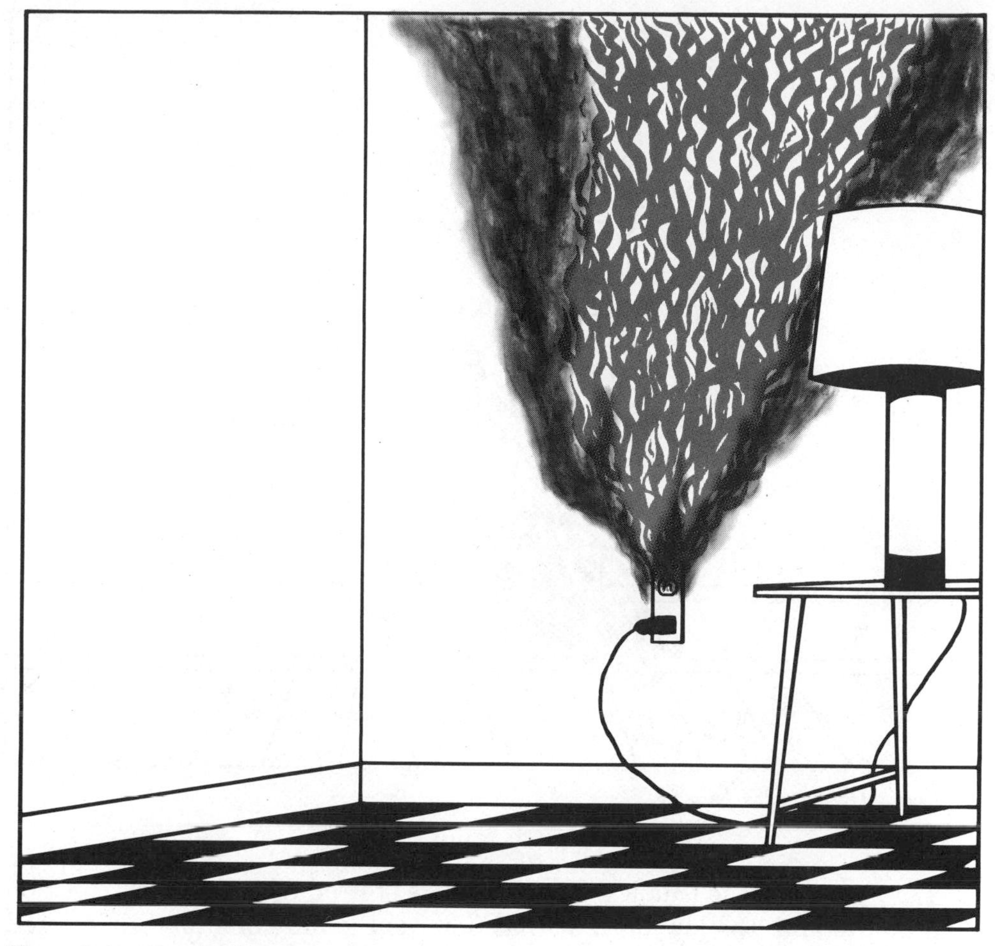

Figure 3-17. *True low burn.*

burn. Then the false low burn may show deeper charring; or there may be more wall and ceiling damage at the false low burn than at the true low burn. Again, the smoke and burn patterns that connect the two low burns can be traced to establish the true low burn. As always, the true low burn must be associated with an ignition source. In addition, occupants or witnesses may indicate—even approximately—where the fire started; this information could be of help in determining which is the true low burn.

Numerous other situations may result in burn patterns that can deceive the eye. The investigator must be very thorough in his examination of the fire damage. The search for the true low burn should not end when one low burn is found.

Example. Figure 3-20 shows three low burns produced by a fire. The shape of each of the V patterns are shown in gray. Before reading further, try to determine which low burn is the true low burn and why.

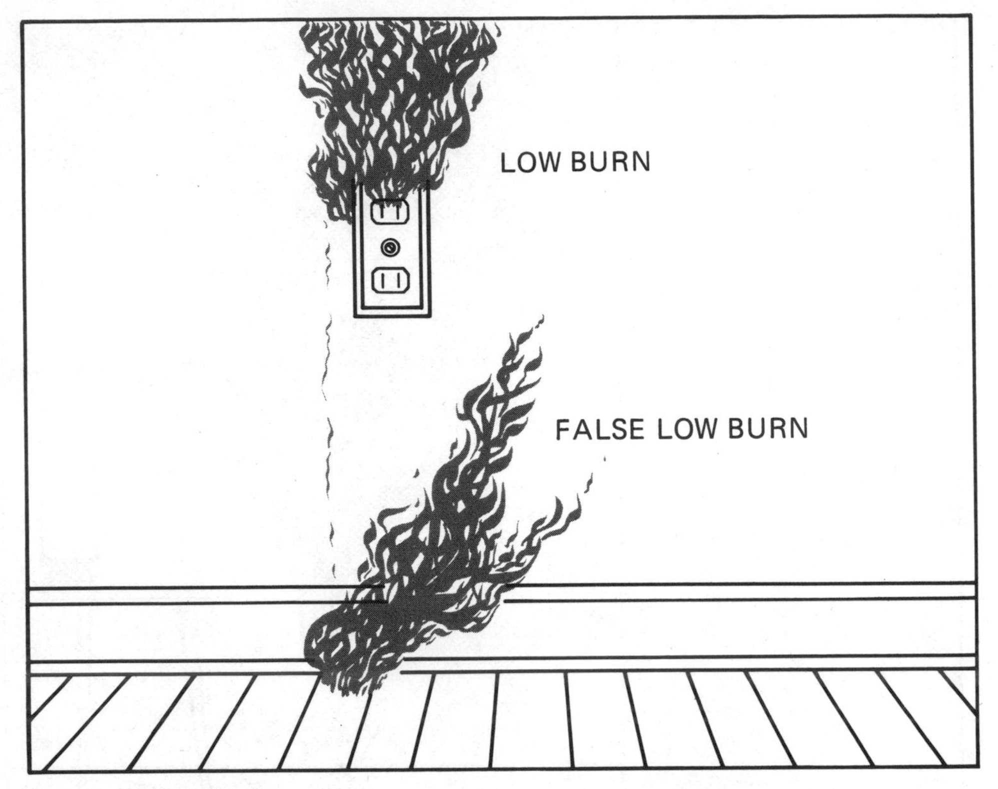

Figure 3-18. False low burn.

Low burn 3 is the true low burn. The fire started in the electric clock on the nightstand; the low burn at that point is the only one directly related to a heat source. Low burns 1 and 2 resulted when flaming curtain material dropped to the floor. Both these low burns are *below* the wall outlet, which is the only heat source near them. Since fire travels up and laterally, the wall outlet could not be responsible for the fire.

Point of Origin

The point of origin is the location of the first material that was ignited. It is closely related to the true low burn and heat source. In fact, once the true low burn is found, the determination of the point of origin is almost routine. The investigator must not, however, arbitrarily pick a point of origin that seems reasonable. The location reported as the point of origin must be the one indicated by the available evidence.

In some situations, where initial flame production is massive, a single point of origin cannot be established. Instead, an area of origin must be designated. For example, a drum of flammable liquid, dropped from a fork lift, might be ruptured when it hits the floor. The liquid would spill over a fairly large area and might be ignited by any of several heat sources. Upon

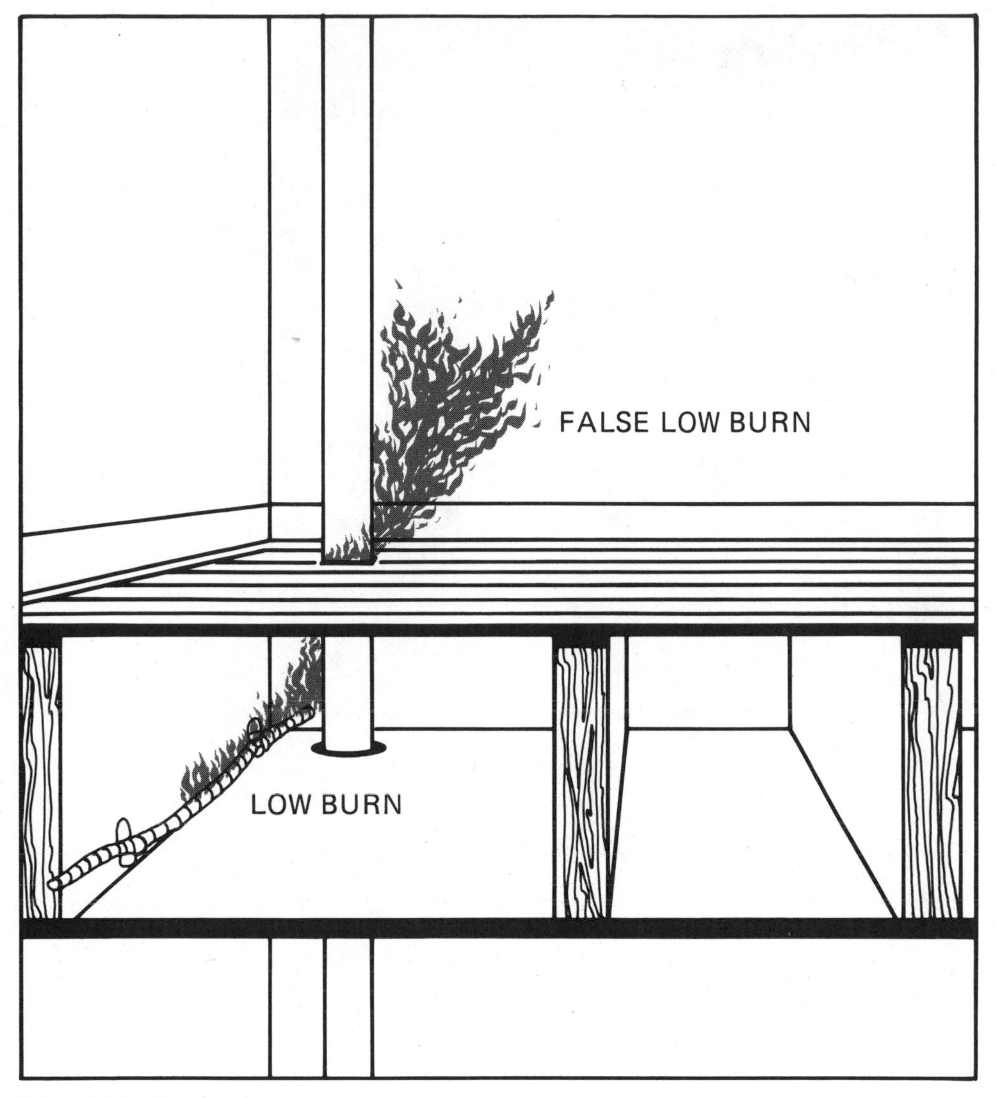

Figure 3-19. True low burn.

ignition, the entire spill would almost immediately be covered with flames. No one point would have ignited first, and no point of origin could be reported. Instead, the entire spill would be reported as the *area* of origin.

Similarly, when accelerants are poured about a room and ignited, the result is an area of origin. The investigator would probably find burn patterns of equal intensity distributed around the room. No single point of origin could or should be reported—but an arson investigation should be initiated as soon as arson seems to be a possibility.

Several examples of points of origin have been given in the case histories so far; many more (including areas of origin) will be given in the remaining chapters.

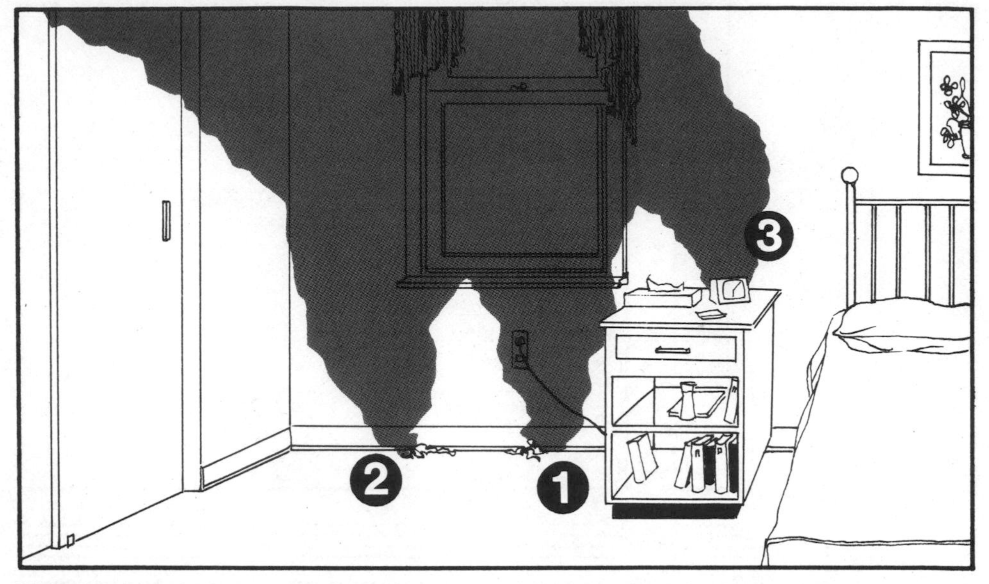

Figure 3-20. *Identify the true low burn.*

4

HEAT (IGNITION) SOURCES

The identification of the ignition source that initiated a fire is one of the four basic objectives of fireground investigation. This information is also important in fulfilling the remaining objectives. First, the true low burn must be directly related to an ignition source in the process of determining the point of origin. Then, the point of origin and the ignition source provide a basis for reconstructing the start of the fire and determining the reason and category.

In this chapter we discuss sources of heat—ignition sources that can and have started fires. Several of these heat sources may be found in almost any structure. Normally, only one is responsible for initiating a fire; the investigator's job is to find out which one. The necessary proximity of the low burn to the heat source is of help in this regard. Where two possible ignition sources are close together, careful study of the burn patterns and the position of the low burn is also of help. In many cases, the investigator can tell from the condition of a heat source whether or not it was responsible for initiating the fire.

MECHANICALLY PRODUCED HEAT

Mechanically produced heat (or simply mechanical heat) is the heat of *friction*. It is produced whenever two surfaces move against each other. For example, when you rub the palms of your hands together, they get warm because the rubbing produces mechanical heat. The friction, which is actually a resistance to the movement, transforms energy from your hands into heat energy.

Sources of Mechanical Heat

Motor shafts are designed to rotate on roller or ball bearings to reduce friction and the heat it produces; lubrication is also used to reduce friction. Drive belts are designed to operate at a specified tension, so they move with their pulleys and thus minimize friction. However, if a motor bearing is faulty or improperly lubricated, friction between the shaft and the bearings can produce enough heat to ignite combustibles—including parts of the motor.

If a drive belt is too loose it can slip and rub against the pulleys; again, this can produce enough mechanical heat to start a fire. If a drive belt is too tight it can rub against the drive pulley, producing frictional heat. Or it can bind the motor shaft, causing the motor to overheat and ignite.

Motors and belt-pulley drives are the two major sources of mechanical heat in sufficient quantities to cause fires. Friction can also be an indirect cause of an electrical fire. If two wires are located so that they rub together, over a period of time their insulation will wear away. Eventually the bared wires will touch, causing a short circuit that can generate enough heat to start a fire.

Any piece of equipment that has moving parts will produce mechanical heat. This includes all motorized equipment, including industrial conveyor systems. (Conveyor systems can also carry fire to uninvolved parts of a structure if they are operating at the time of a fire.) Lathes, milling machines, metal stamping machines, grinders, and other machine tools also produce mechanical heat where the tool meets the work piece.

In dwellings, refrigerators contain motors and usually drive belts, and thus can produce mechanical heat. Duct blowers, ventilation fans, air-conditioner and humidifier motors, and even the motors of record-player turntables can and have caused fires by generating mechanical heat. In many cases, these units are overlooked as sources of mechanical heat, and their electrical components are blamed for the fire.

Investigative Aspects

When a drive belt is suspected of being the heat source for a fire, examine the remains of the belt. Burning should be evident on the inside of the belt, the side that touches the pulley. The pulley itself should show scorching or burn patterns. The low burn should be in the vicinity of the belt. However, the belt can burn through and fall to the floor, causing a low burn below its operating position. Fraying may also indicate that the belt was operating improperly. In some instances, the belt will fray to the extent that it tears itself apart.

Ask occupants in a dwelling, or employees in a factory, if the equipment was operating properly. Such responses as "The belt squeaked a lot" and "The belt started smoking just before the fire" are obviously of help in finding the heat source. Occasionally, an occupant or worker will note that he or she saw the belt burst into flames. This should be checked by conducting a thorough examination.

In motors, mechanical heat is most often produced at the bearings. Inspect the bearing housing. A burn pattern extending out from the faulty bearing should be visible. Before deciding that mechanical heat caused a fire, make sure that enough heat could have been generated to ignite nearby combustibles. Then, through questioning, determine whether the source of mechanical heat was at its normal location.

Shock

A shock, such as the blow of a hammer on a nail, is a mechanical process. It can produce heat through friction. A drum falling off a hi-low may skid along a concrete floor, producing a hot spark. If the drum contains a flammable liquid, and the fall ruptures the drum, the spark may ignite the liquid. If a railroad tank car is involved in a collision, the shock may rupture the tank; at the same time, the impact can produce enough frictional heat to ignite the spilling contents of the tank.

Shock alone can cause the explosion of an unstable compound—an explosive, for example, or an acetylene tank. The disruptive force of a shock can also cause damage that eventually leads to a fire. The vibration of a jack hammer at an excavation site can disrupt underground gas lines sufficiently to cause a leak; staples or nails that are fired by gun-type construction tools can disrupt or pierce electric circuits or piping, and thus indirectly cause a fire.

Investigators rarely consider shock as the possible cause of a fire because it is not very common and because it is difficult to detect and prove. The momentum of colliding bodies carries them away from the collision site; explosions destroy evidence and scatter debris. It may be extremely difficult to determine whether fire occurred during or immediately after impact. Yet in spite of these difficulties, shock and the resulting frictional heat should come to mind as a possible ignition source when a disruptive force is involved.

SOLAR HEAT

There is only one source of solar heat—the sun. Radiant heat from the sun can be focused onto combustible materials by a magnifying glass, window pane, bottle, a glass of water, or almost any other transparent object (Figure 4-1). In time, the combustible material will ignite. For example, over a period of days, a wooden window sill can be pyrolyzed by sunlight concentrated through a distorted or imperfect window pane. This same solar heat can ignite vapors from the window sill, causing a fire. A flimsy material, such as a sheet of paper, may be ignited shortly after it is exposed to

Figure 4-1. *Sun rays passed through glass to char window frame.*

the concentrated solar energy. Outdoors, leaves and dry grass can be ignited by sunlight passing through a piece of broken glass or an old bottle.

Presently, systems are being installed on structures to capture solar energy and use it for space and hot-water heating. A major problem with such installations is in controlling the captured heat. If it is overly concentrated in a particular area, it can ignite the contents in that area.

Investigative Aspects

If the investigation of a fire eliminates all other heat sources, if there is a window in the general area, and if the fire occurred on a sunny day, solar heat should be considered as the ignition mechanism. Determine the approximate time of ignition (from the fire reports). Then determine the angle at which the sun would have entered the window at that time. Check for the remains of a bottle or other glass object that may have concentrated its rays. If the sun's angle coincides with the line-of-sight angle between the window and the point of origin, the sun was probably the heat source.

CASE HISTORY: Solar Ignition

On a sunny July afternoon, firefighters responded to a basement fire in a two-story dwelling. The fire was in the free-burning stage, producing gray-white smoke. Firefighters were able to gain entry and extinguish the fire quickly.

The fireground officer immediately began an investigation. The smoke and heat patterns led him to a table standing against the basement wall, near a washing machine. The fire had obviously started at the top of the table, which held the remains of clothing and cardboard soap-powder boxes. The fire had formed a wide V pattern up the basement wall and had burned across the ceiling toward a stairway.

The officer had difficulty finding a heat source that could have caused ignition. The switches of the washing machine and a clothes dryer were in the off position. The female occupant confirmed that the washer and dryer had not been in use. There were no other electric cords or devices in the area, and the electric outlet was obviously not involved. Moreover, all the occupants of the dwelling were nonsmokers.

An investigator from the Fire/Arson Investigation Squad was sent to the scene. He noticed a clear, cut-glass decanter standing on the basement window sill, directly above the wash table. He immediately checked the weather conditions and the position of the sun prior to the fire. This information confirmed his assumption: The sun's rays, shining on the window, had been focused onto the table top by the bottle; the heat had ignited the combustibles there. His determination was as follows:

Point of origin: Top of the wash table
Heat source: Solar energy
Reason: Sun's rays concentrated through a bottle
Category: Accidental.

CHEMICALLY PRODUCED HEAT

In a chemical reaction, two or more substances combine (or react) to form new substances. Some chemical reactions require the addition of heat in order to take place. Other reactions give off heat when they occur. In the latter case, enough heat may be produced to ignite nearby combustibles. In addition, if any of the original substances or reaction products are combustible, they too may be ignited.

Chemical reactions that produce heat are not limited to laboratories or complex chemical industries. Many common substances undergo these reactions.

Spontaneous Heating and Ignition

A number of common materials are subject to spontaneous heating and ignition. These include natural oils such as vegetable and fish oils (but not hydrocarbon-base oils), some types of soft coal and charcoal, and cut grass and hay. These materials react slowly with oxygen or water, producing heat. (The reaction is much like an extremely slow combustion.) If the oxygen supply is not limited—for example, by storing the material in a closed metal container—and if the area is not ventilated to remove the heat, enough heat will accumulate to ignite the material.

Oily rags, left in a calm, open space, are especially susceptible to spontaneous heating and ignition; the oil reacts with oxygen to produce heat, which then ignites the combustible rags. Green grass, placed in a container in a warm area, will create sufficient heat to ignite itself (and the container if it is combustible).

Hay that is baled and stored before it is properly cured (dried) is subject to spontaneous heating and ignition. Here the chemical reaction occurs between the hay and the water that remains in the bale. If sufficient oxygen is available, the heat produced by the reaction will ignite the hay; then the stack of bales and often the barn will become involved. Enough heat for ignition may be produced within a pile of bales, but tight packing of the bales may keep oxygen away from the fuel. As soon as a few bales are removed from the pile, however, oxygen is able to reach the hot inner bales and ignition results. Occasionally, when a bale of hay or cotton is opened, the center is found to have burned black. This indicates that there was enough oxygen available to allow ignition of the hot center, but not enough to sustain burning.

Noncompatible Substances

Pine oil and dry chlorine bleach are common household chemicals. Pine oil is used to deodorize bathrooms, basements, pet bedding, and so on. Dry chlorine bleach is an ingredient of many products used to wash clothing. In many homes, all cleaning products are stored in the same area. Yet, when pine oil and dry chlorine bleach are mixed, they react spontaneously. Heat is produced, along with chlorine—a white, very toxic, highly flammable gas. The heat of the reaction is sufficient to ignite the chlorine, which burns with a red flame.

Commercial and Industrial Chemicals

Many of the chemicals normally found in commercial and industrial buildings will react violently with each other. The metals potassium and sodium react with water, producing enough heat to self-ignite. A number of metal powders can ignite on contact with moist air.

The available chemicals number in the thousands, and the combinations of chemicals in the millions. It is impossible to remember all the combinations that can produce fire or an explosion. When a fire involves industrial chemicals, the fireground investigator should request the assistance of a chemist.

Investigative Aspects

Before concluding that ignition was produced by spontaneous heating or some other chemical reaction, look for more common heat sources near the low burn. Make sure that chemical heat might be a possible ignition source in the given circumstances. For example, only certain materials are subject to spontaneous heating. Look for the remains of those materials—hay in a barn, soft coal, oily rags, cut grass. Check remaining bales of hay when investigating a barn fire. Check with occupants or owners concerning the presence of these materials.

Chlorine gas, the product of a pine oil and dry bleach reaction, has a very irritating odor and produces red flames. Question witnesses about flame color and smell. Determine from occupants whether these two chemicals were used in the building, and where they were stored. Check the area of the low burn for remnants of the chemicals.

If commercial or industrial chemicals are found and there is no other logical ignition source, carefully collect any containers that were involved in the fire. Keep the containers separated by placing each in a cardboard box by itself. Have their contents analyzed to determine whether they could have reacted and produced sufficient heat for ignition.

Treat the containers as evidence, by maintaining their integrity and the chain of custody (Chapter 8); the chemicals may have been employed by a clever arsonist to disguise the use of flammable liquids. Try to determine, through questioning, which chemicals (if any) were part of the normal contents of the fire structure. Again, do not hesitate to request technical help if it is needed.

CASE HISTORY: Noncompatible Substances

A destructive supermarket fire was found by investigators to have resulted when a bottle of pine oil was broken, and the oil spilled down onto a shelf containing boxes of dry laundry powder. Because the laundry powder contained only a small amount of chlorine, the pine oil-chlorine bleach reaction produced heat very slowly. If the spill had been cleaned up during shopping hours, a fire would have been avoided. However, the spill was ignored and the resulting fire was not discovered until after 3 A.M. the next day. By that time, the supermarket interior was almost completely involved via flashover, and the steel-truss roof over the point of origin was sagging. The investigators' report read as follows:

Point of origin: Shelving containing pine oil and dry laundry compound
Heat source: Chemical
Reason: Incompatible household chemicals mixed and reacted, creating heat
Category: Accidental.

ELECTRICALLY PRODUCED HEAT

Heat produced by electrical wiring, appliances, and equipment is responsible for many fires. But many other fires are incorrectly blamed on electrical components. Sometimes a mistake is made because complex electrical equipment or wiring is involved, and the fireground investigator does not have the knowledge to make an accurate judgment. In other cases, an electrical component that has become involved in the fire is automatically labeled as the heat source.

Fireground investigators are expected to learn as much as they can about the structures, equipment, processes, and materials that may become involved with fire. However, they are not expected to know everything. Residential wiring can be complex and industrial equipment and electrical systems are often close to impossible to decipher. If there is any doubt about the involvement of electrical components, the investigator *must* request a technical investigation to utilize the knowledge and experience of experts.

Electrical Wiring

Electric wiring is frequently involved in fire, mainly because there is so much of it in most structures. When wiring is the heat source for a fire, heat produced within the wiring moves out through the insulation to ignite nearby combustibles. When wiring is involved but is not the heat source, it is attacked by the fire from the outside in. These two situations produce two different kinds of damage that may be distinguished by the investigator.

Indications of Attack by Fire. When wiring is attacked by fire, the insulation is melted or burned away. As the melted insulation cools, it becomes brittle and adheres to the wire; the wire cannot be rotated within the insulation (Figure 4-2).

When the insulation is burned away, the copper wire is thinner than normal. This is due to heating and subsequent stretching or sagging of the copper—from its own weight or the weight of, say, a junction box that might have dropped from the involved wall. Beads of copper may be observed on the wire (Figure 4-3) if the fire was particularly intense. Copper will flow at a temperature of about 1980°F (1082°C) and forms beads as it cools.

The copper wire may break because of the stress placed on it by the heat and falling objects. The broken ends are then jagged and rough (Figure 4-4).

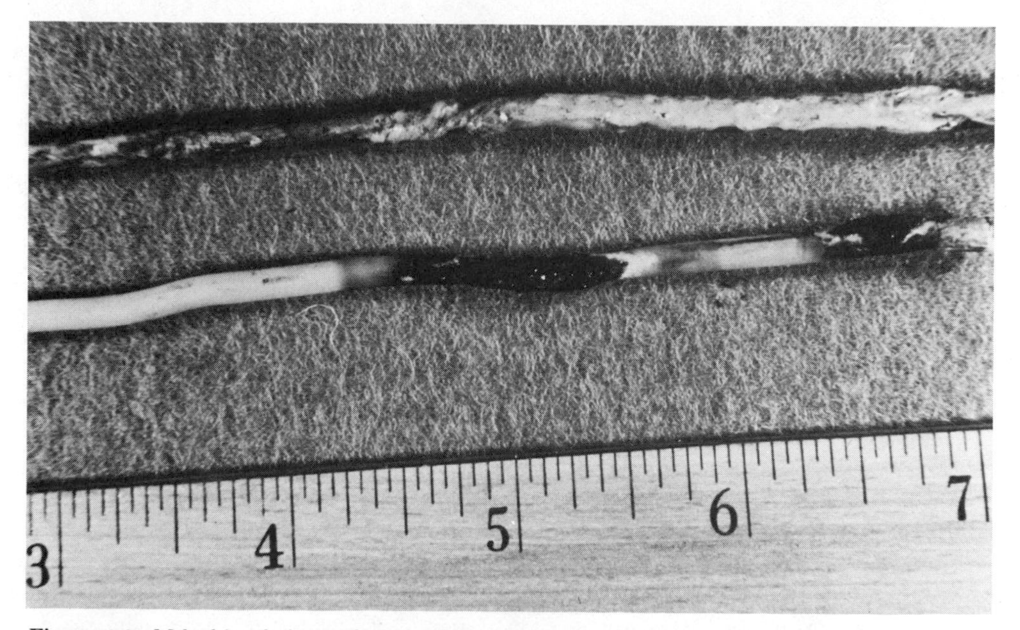

Figure 4-2. *Melted insulation cools and adheres to wire.*

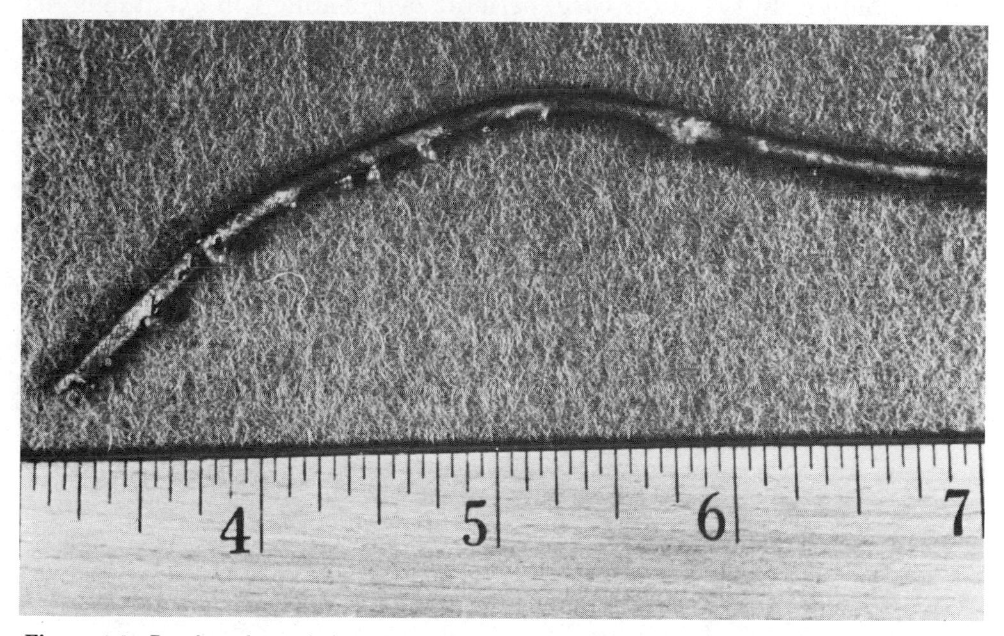

Figure 4-3. *Droplets along wire's surface form following copper flow.*

Indications of Internal Heating. Wiring that is carefully installed according to building codes will produce a small amount of heat. However, the wiring will overheat sufficiently to start a fire when it is damaged, overloaded, the incorrect size wire or fuse is used, or a short circuit occurs. The damage might be caused by a sharp bend or a pinch created when the wire was installed. A surge of electricity, as from a lightning strike, or the stress of

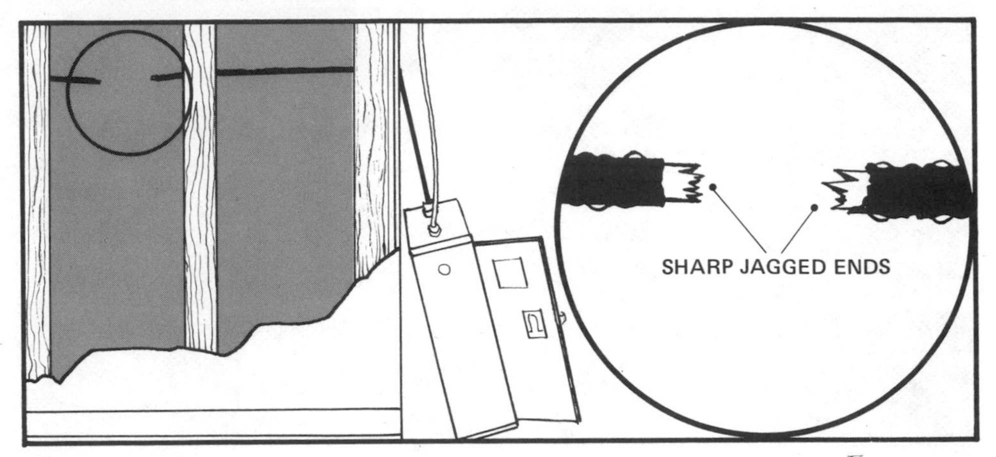

Figure 4-4. Jagged ends of copper wire are caused by stress from heat and falling objects.

a tornado or earthquake, might do some damage but not enough to break the wire or cause an immediate fire. At some later time, overheating may occur in the damaged area. The overheating may start a fire or just damage the insulation. In the latter case, periodic overheating will eventually dry out the insulation and make ignition much more likely.

When wiring is damaged by internal heating, the insulation is burned from the inside. It is then loose, more like a sleeve than insulation, and will easily slide back and forth on the copper wire. Bubbles are visible on the insulation (Figure 4-5). Unless there is a loose connection at a junction box, the wiring will have overheated along its entire length; the signs of over-

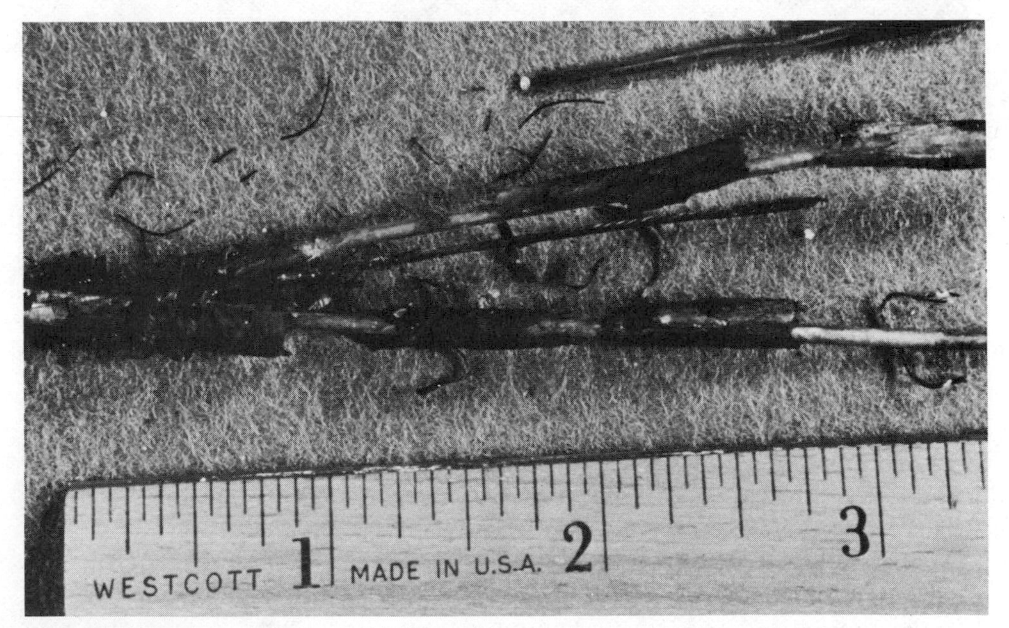

Figure 4-5. Loose insulation caused by internal heating.

heating should be visible all the way back to the circuit breaker or fuse box. If there is a loose connection in the wiring, it should show signs of arcing.

The intense heat within the wiring causes the copper to form bubbles. The bubbles burst and blow to the outside of the wire, forming small balls of copper (Figure 4-6) and leaving craters and rough spots on the wire (Figure 4-7). The two wires may be fused together; the investigator should strip some insulation and check for this condition (Figure 4-8).

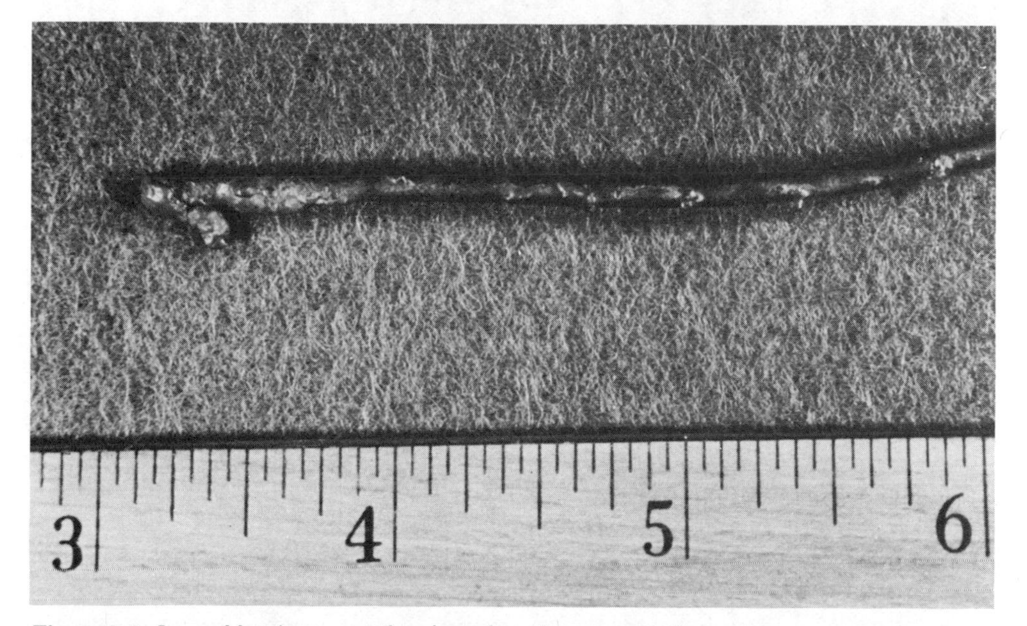

Figure 4-6. *Internal heating causes the wire to form tiny, copper balls.*

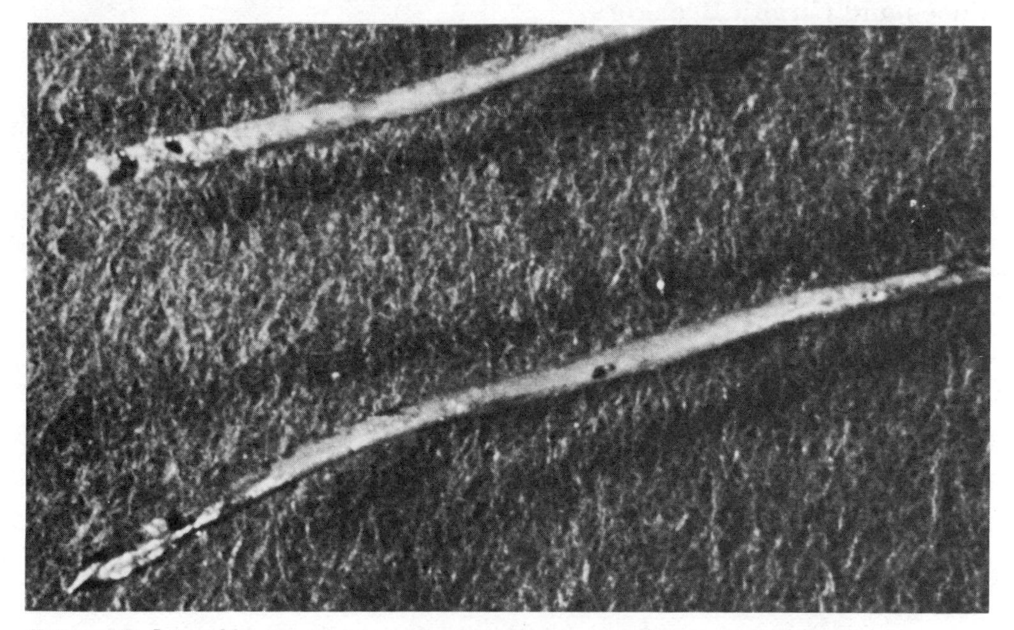

Figure 4-7. *Internal heating causes rough spots and craters on wires.*

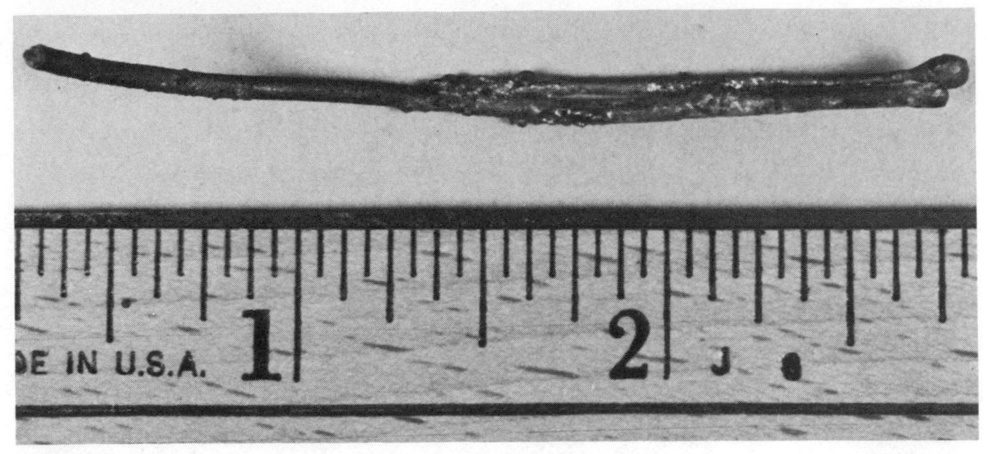

Figure 4-8. *Fusing of wires caused by internal heating.*

Copper wires that have lost their insulation are discolored by the heat. The discoloration varies from orange to red, with red indicating exposure to the greatest amount of heat.

The heat generated within the wiring may melt through the copper wires. In some cases there will be beads of copper on the ends of the broken wires (Figure 4-9). Or, arcing across the wires may cause copper particles to splatter on surrounding materials, where they may be found by the investigator. If a piece of the wire is missing, it has probably melted, fallen, and then solidified as small copper balls of various sizes. These may be recovered from the debris below the wire (Figure 4-10).

Fuses and Circuit Breakers

Fuses and circuit breakers are designed to open electric circuits so as to prevent overheating of electric wiring and components. It is therefore

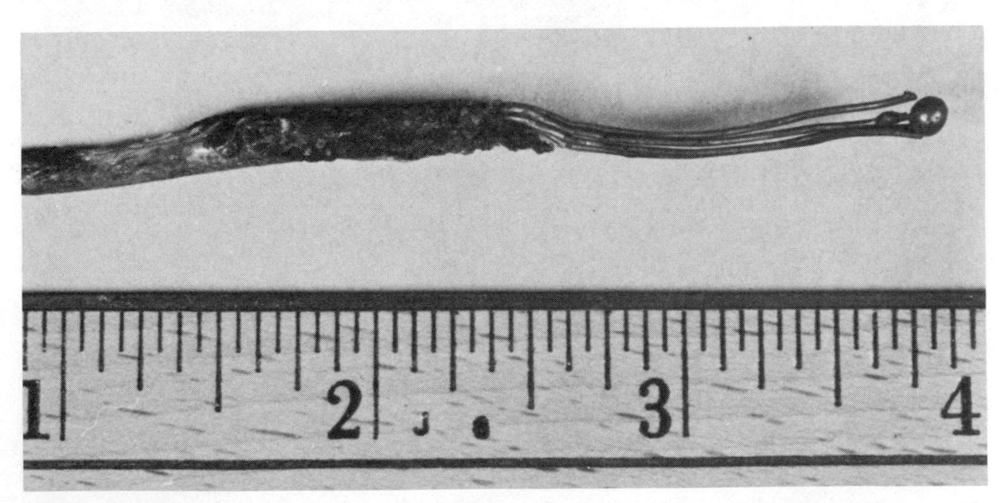

Figure 4-9. *Internal heating causes copper balls on ends of wires.*

Figure 4-10. *Missing pieces of wire could solidify as small copper balls.*

frustrating to find that a fire was caused by electrically generated heat when the fuse or circuit breaker did not operate properly. Some of the reasons for improper operation are:

- Overfusing, the use of a fuse or circuit breaker that is rated too high for the application. For example, a 30-ampere fuse in a circuit requiring a 15- or 20-ampere fuse will provide little, if any, protection.
- Using a coin in place of a fuse. This is sometimes done to keep a faulty appliance from continually burning out fuses. The coin provides no protection at all.
- Installing the fuse or circuit breaker box on a cold wall. This keeps the circuit breaker cool during electrical overload, when it should be getting hot enough to operate.
- Failure of the protection device itself.

Investigative Aspects

Inspection of fuses and circuit breakers can sometimes reveal when a short circuit has occurred in the electrical system. A short circuit is indicated when the face of a fuse is black and the fusible metal bar is broken. The brass screw threads of the fuse may be discolored or even burned away (Figure 4-11). Check the bottom of the fuse and the fuse socket for discoloration or melting (Figure 4-12). Circuit breakers have been known to melt inside the breaker box without tripping; check for this condition.

The fuse or circuit breaker box may itself be the point of origin; if so, it will show characteristic heat or burn marks. If the door of the box was

Figure 4-11. *Discolored or burned away brass screw threads indicate a short circuit.*

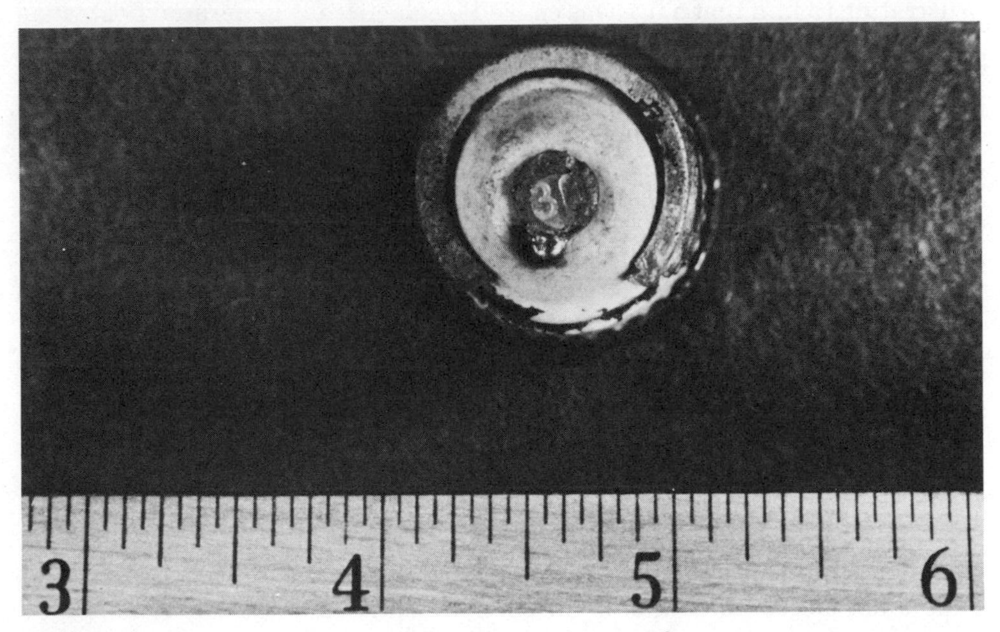

Figure 4-12. *Short circuits often cause the bottom of the fuse to melt.*

closed, the marks will extend outward from cracks around the perimeter of the door. If the door was open, the marks will extend upward on the inner surfaces of the box and the door (Figure 4-13) and perhaps on the wall behind the box.

Figure 4-13. *If the fire started in the fuse box, the burn pattern will extend out and up the walls.*

CASE HISTORY: Tampered Fuse

The fire department responded to and extinguished a basement fire in a single-family dwelling. Firefighters, when questioned, reported seeing only the gray smoke indicative of ordinary combustibles. One investigator followed the smoke and burn patterns to a low burn at the basement fuse box. This was also the area of deepest charring. Patterns on and around the fuse box indicated that flames emanating from the box had ignited clothing that was hanging nearby.

The fuses were removed from the box and examined. On one fuse, the thin brass screw-thread cap had been cut to form a flap. The flap had been bent down under the fuse, and the fuse replaced. This completed the circuit, so the fuse could not fail. The circuit was unprotected.

A second investigator was questioning the occupants. They had been using several appliances prior to the fire and had heard a buzzing sound just before some of the appliances stopped functioning. Soon after that, they smelled smoke and discovered fire in the basement.

On further questioning the occupants noted that several days earlier, insulation installers had had a problem with fuses blowing while they worked. Investigators were told the workmen were the last people to touch the fuse box. This was checked and found to be true. The workmen had shorted the fuse so as to work without interruption. The investigators' findings were:

Point of origin: Basement fuse box
Heat source: Overheated fuse and socket
Reason: Overfused circuit (tampered fuse)
Category: Accidental.

Electric Cords

The electric cords for lamps and small appliances contain light-duty wire and minimal insulation. They can overheat and cause a fire without blowing a fuse or tripping a circuit breaker. Both lamp cords and light-duty extension cords may become frayed—and then cause a short circuit—if they are placed under rugs and walked on continually.

In addition, these light-duty cords may become overheated when they are located under rugs. They are designed to be used in the open and cooled by normal air circulation. However, there is no circulation under a rug; the heat produced by the cords is trapped and cannot dissipate, and the cords can overheat and cause ignition. This also may happen when an extension cord is located in a confined space, such as a clothes closet.

Light-duty extension cords may overheat when they are used with heavy-duty appliances or tools. Then, the heavy current is simply too much for the light wiring to carry. In some cases, animals have caused short circuits by chewing through light-duty electric cords.

Overheated electric cords produce the same indications of internal heating as does wiring (page 75).

Small Appliances

Small electrical appliances are responsible for many fires—because there are so many of them, they are used frequently, and they are often used carelessly. Worn mechanisms and worn electric cords can cause overheating. Flames may be produced within the appliance itself. In some cases, the user may unplug the appliance and move it before flame production actually begins. Then the unplugged appliance may not seem to be a logical source of ignition. However, fire most often results when a hot electrical appliance contacts surrounding combustibles.

Because small appliances are portable they may be used anywhere if an electric outlet is available. And many have been used in strange places! The investigator must depend on common sense in deciding whether an unusual location is, nevertheless, a reasonable one. His first objective is to locate the point of origin and heat source and to connect them. Once that is accomplished he can try to determine why an appliance may be in an unusual place. If the explanation is not reasonable, the fire may be termed suspicious; then arson investigators (or the courts) will make the final determination.

Small appliances that can produce enough heat to start a fire are those that contain resistance-type heating elements. Generally, they can cause ignition in three ways: First, a normally operating appliance may be placed too close to combustible materials (and perhaps left unattended); the normal operating temperature is sufficient to ignite the combustibles. Second, some mechanism within the appliance (usually the internal thermostat) may fail to operate properly; excessive heat and perhaps flames produced by the appliance will ignite combustibles that would normally be at a safe distance. Third, the electric cord may become damaged and cause a short circuit; often, it is the appliance itself that burns or melts the insulation on a carelessly placed electric cord.

In addition, some types of small electrical appliances are prone to particular fire hazards.

Toasters. The outer case of a toaster can become hot enough, in normal operation, to ignite combustibles on contact. If the bread-release mechanism fails to operate, the bread may ignite in the toaster, producing flames that can travel to nearby fuels.

Irons. Irons are very often left unattended while they are hot. They will easily ignite combustibles such as ironing-board covers and clothing, and quickly burn through electric cord insulation.

Hair Driers and Hair Curlers. Failure of the internal thermostat in a hair drier can cause sufficient heat production to melt the appliance and ignite surrounding combustibles. Failure of the blower allows a tremendous amount of heat to build up in and around the hair drier. The combination of a hot hair drier and flammable hair spray can result in ignition, as can misuse of the appliance to dry clothing or rugs.

Hair curlers are subject to the same hazards as hair driers (except for blower failure). They are often placed too close to combustibles while hot.

Portable Electric Heaters. These devices can ignite nearby combustibles by radiation as well as through direct contact. Failure of the internal fan can cause overheating.

Hot Plates. The heating element of a hot plate is exposed and thus will ignite combustibles that may be dropped onto it. Heat radiated from the element can cause ignition at a distance. Like irons, hot plates seem often to be left unattended while operating.

Coffee Makers. An electric coffee pot can become extremely hot when the coffee is allowed to boil off. Another problem is careless moving and placing of stock near operating coffee makers in commercial establishments.

Electric Blankets. These appliances rarely cause fire, unless they are misused. Overheating can result when an electric blanket is operated while it is folded, with other blankets on top of it, or even when the user of the blanket lies on it.

CASE HISTORY: Electric Blanket

A mid-afternoon fire involved several rooms in a rooming house. After extinguishment, the fireground investigator traced the smoke and burn patterns to the bed in a corner room. The occupant of the room was at work, but other building residents noted that the occupant had the habit of smoking in bed. They felt sure that was the cause of the fire. However, the low burn was at the foot of the bed. The investigator removed the remains of a heavy quilt from the bed and found an electric blanket under it. The blanket was folded at the foot of the bed; its electric cord was plugged into a wall outlet, and its control knob was set at "on."

The occupant was questioned when he returned to the rooming house. He said his morning routine included turning off the electric blanket, folding it at the foot of the bed, and spreading his quilt over it. Although he was sure he had turned the blanket off that morning, he obviously had not. The investigator found:

Point of origin: Foot of the bed
Heat source: Electric blanket
Reason: Improper operation, causing overheating of the blanket
Category: Accidental.

Electric Motors

Earlier in this chapter, we noted that motors may cause ignition by producing heat mechanically. The electrical components of a motor may also produce enough heat to ignite the motor itself and nearby combustibles. This happens when the armature (rotor) short circuits and fuses to the field (or case); then the electrical energy entering the motor is transformed into heat rather than motion.

Investigative Aspects. Check the motor's shaft. If it can be rotated, there was no fusing or short-circuiting; burning did not originate in the motor. If the shaft is frozen and cannot be rotated, heat was most likely produced within the motor. Check for fusing of the internal windings (wires) and for melting of the armature or fusing to the motor casing.

Major Appliances

Smoke and burn patterns can easily be traced back to a major appliance that was the source of heat for a fire. However, it may be much more difficult (or perhaps impossible) for fireground investigators to determine exactly what happened within the appliance. Any one of many different mechanical or electrical components and systems might have been the heat source. If a lawsuit develops as a result of the fire, it is simply not sufficient for the investigator to note that "the fire started in the refrigerator." More important, if the investigator doesn't know exactly where in an appliance the fire originated, and what exactly was the heat source, his determination of a category is, at best, a guess.

In certain situations, the investigator can be sure of the heat source. Grease, food, or combustibles that are ignited by an operating burner on an electric range are obvious examples. Newspapers placed on top of an operating television set, blocking the heat vents and being ignited by the heat buildup, may not be so obvious; but this heat source may be discovered by an experienced investigator. However, ignition in the wiring of a kitchen range or ignition resulting from the "instant-on" capability of some television sets will not be easy to detect, even for an experienced investigator.

Appliance repair personnel are quite familiar with the malfunctions of these units, as well as their operation. Local service personnel or factory representatives can probably find the exact heat source quickly and accurately. Expert help should be requested if at all necessary. If the work cannot be performed at the fireground, the appliance should be moved to a fire department facility for examination. (If any property is removed from a fire scene, the owner must be given a receipt, as outlined in Chapter 8.)

Light Bulbs and Fixtures

Incandescent light bulbs operate at a high temperature; they can ignite such combustibles as clothing and paper on contact. Like light-duty electric cords, bulbs are meant to be cooled by normal air circulation. In a poorly ventilated recessed light fixture, heat from the bulb can deteriorate the wiring insulation, causing a short circuit and fire. The heat buildup can also ignite nearby combustibles. In a closed space such as a closet, with the bulb operating, enough heat can accumulate to ignite stored combustibles.

Fluorescent light tubes do not produce enough heat to cause ignition, but the starters (ballasts) in fluorescent fixtures can become short-circuited and then produce sufficient heat for ignition.

Investigative Aspects. If an incandescent light bulb seems to be the heat source, look for evidence of direct contact with flammable materials. If the point of origin was not disturbed by firefighting operations, the bulb may still be in contact with combustibles. Severely charred fragments of the original fuel may be stuck to the bulb, or to bits of bulb glass on the floor below the light fixture. The charred fragments may be analyzed and identified at a laboratory if necessary.

CASE HISTORY: Light Bulb Fire

A fire involving the entire second story of a two-story single-family dwelling was extinguished after some difficulty in venting the smoke. Smoke and burn patterns led investigators to a clothes closet in a second-story bedroom. Inside the closet, some clothing had been completely burned off the hangers; other clothing had fallen to the floor. Copper wires hung down from the ceiling where a light fixture obviously had been located. The wires showed no sign of internal heating, nor did the fuse box.

Three children, but no adults, had been at home at the time of the fire. They denied playing with matches or smoking in the closet. The eldest of the three said she had looked for a swimsuit on a closet shelf some time before the fire. She admitted that she may have forgotten to turn the light off.

Careful digging through the debris on the closet floor produced the remains of a hanging light fixture—again showing no signs of internal heating. No charring was evident on the closet floor when it was uncovered. However, a piece of light bulb glass about the size of a half dollar was found. Stuck to the glass was charred but recognizable cloth.

The evidence and the children's statements led the investigator to the following determination:

Point of origin: Closet light fixture (hanging)
Heat source: Light bulb
Reason: Clothing was pushed against operating light bulb
Category: Accidental.

GAS SERVICE AND APPLIANCES

Most gas-burning units—ranges, heating plants, and industrial process burners—provide a necessary ignition source in the form of a pilot flame. When heat is demanded, gaseous fuel flows to the pilot light, where it is ignited. When sufficient heat has been produced, the flow of gaseous fuel stops, but the pilot flame continues to burn. Most often, hostile fire results when combustibles are placed too close to the unit. However, the pilot flame may be an inviting heat source for an arsonist.

Gas meters and gas piping are sources of fuel but not sources of ignition. No heat is generated in gas service equipment, even when gas is leaking. The gas must reach a source of heat (or one must reach the gas) for ignition or explosion. Gas lines or meters may leak or rupture because of corrosion, improper installation, collapse of structural elements during a fire, or expansion or melting of pipes and pipe joints during a fire. When a fire causes a gas leak, the gas will feed the fire; but the gas leak is not the point of origin, heat source, or reason for the fire.

Investigative Aspects

Do not jump to a quick conclusion if nearby combustibles are ignited by a gas-fired appliance. Make sure the combustibles were in their normal position or were accidentally placed near the appliance. Thoroughly examine the burned materials at the point of origin for residues of flammable or combustible liquids. An arsonist may have provided for delayed ignition via the appliance's thermostat, and used accelerants to ensure rapid burning.

If gaseous fuel was involved in the fire, try to determine whether the fire caused gas to leak or whether leaking gas found an ignition source and started the fire. Remember that leaking gas will move through a building. Consider the burn patterns and low burn, the positions of gas pipes, whether manual gas controls are in the "on" or "off" position, and whether automatic controls are operative.

Check on and around controls for wrench marks or damage that may have prevented the controls from operating normally. Pay particular attention to the gas meter in this regard. Check the piping for missing threaded plugs or caps or other signs of tampering.

HEATING AND HOT-WATER UNITS

These devices range from simple electric baseboard units for domestic heating to huge, complex industrial boilers. They produce heat and fire in normal operation, and they can be the cause of hostile fire through malfunction, carelessness, or deliberate acts of arson. Because there are so many types and variations of heating units, expert help may be required when one of these units is suspected to be the ignition source for a hostile fire. Here again, the investigator can help himself by learning as much as possible about heating and hot-water devices.

Solid-Fuel Heating Units

Coal- and wood-burning heating units are being used in increasing numbers, in reaction to the high price of heating oil. In many cases, they are installed and maintained by novices rather than heating experts. Then, destructive fires are caused by:

- Installation of the unit too close to combustible walls, or storing combustibles too close to the heating unit
- Improper installation of the chimney, installation of the wrong chimney, or the use of an existing chimney that is not designed for solid-fuel combustion
- Lack of chimney care, especially with wood-burning stoves
- Burning of green wood, which produces a heavy creosote buildup on inner chimney walls, leading to chimney fires

- Improper use of the heating unit controls
- Improper loading of the unit
- Failure of automatic feeding devices (on units so equipped)
- Lack of maintenance of blowers and motors (on units so equipped).

Gaseous-Fuel Heating Units

Gas heating units are fueled by natural, propane, or butane gas. These gases have different physical properties, but the heating units present similar hazards. The gas flow valve may malfunction, allowing the continuous flow of gas, with the pilot light off. The thermostat may fail, allowing overheating. A furnace blower motor may be subject to any of the heat-producing malfunctions discussed earlier in this chapter. A steam or water boiler may explode through failure of the pressure relief valve. And, combustibles may be placed too close to the heating unit.

Oil-Fueled Heating Units

Proper maintenance is the most effective fire-prevention measure for oil-fired heating units. Conversely, lack of maintenance can lead to fire through the failure of controls, motors, or the burner itself. Blower motors, boilers, and nearby combustibles also may cause or be involved in the ignition of a hostile fire.

Electric Heating Units

These are thermostatically controlled resistance-heating units, usually of the baseboard type. Failure of a thermostat can cause sufficient overheating for ignition of wall finishing materials. Blower motors, where provided, may cause ignition by the methods discussed earlier in this chapter.

Hot-Water Heaters

The major hazard involved with hot-water heaters is explosion resulting from overheating and failure of the relief valve. Combustibles placed too close to a water heater may be ignited. In a non-electric water heater, leaking fuel (usually a gaseous fuel) may become ignited and burn or explode.

Investigative Aspects

Most important are the smoke and flame patterns on or around (and sometimes in) the unit. Try to determine whether the unit was the ignition source. Check for nearby combustibles that might have been ignited by heat from the unit. Check thermostats, filters, motors, and drive belts for malfunctions. For gas- and oil-fired units, look for leaks, breaks, or tampering in fuel lines.

Check whether the chimney of a solid fuel unit produced fire by pyrolizing wood framing. Generally, observe whether the chimney could have caused ignition.

If a hot-water heater or a heating-system boiler exploded, check the relief valve and thermostatic controls for proper (or improper) operation.

OPEN FLAME

The most obvious ignition source is an open flame—from a match, cigarette lighter, candle, propane torch, barbecue grill, or the supposedly controlled burning of leaves or other trash. We also include glowing tobacco from cigarettes, cigars, and pipes and, nowadays, carelessly discarded embers from coal and wood stoves. Open flame ignition sources are responsible for large numbers of fires, mostly accidental in nature. However, they are so easy to obtain and use that they are a prime tool of arsonists.

Investigative Aspects

For the most part, open flame should be considered an ignition source when no other logical heat source is available at the point of origin. (Of course, there may be direct evidence of open flame ignition—for example, candle wax or a melted cigarette filter—at the point of origin. However, most open flame heat sources are consumed by the fire they ignite or are carried away.) The investigator who decides that a fire was started by an open flame must then determine whether it was accidental or intentional.

The location of the point of origin is sometimes of help here. Cigarettes or lit matches that are carelessly placed in ashtrays may fall out and ignite combustibles. The point of origin is then in the area of the ashtray. Smokers who fall asleep while smoking usually do so in beds, on couches, or on soft, comfortable chairs. Smokers may throw unextinguished matches in wastebaskets. Hot contents of ashtrays may also be thrown into trash baskets or bags. Children may experiment with matches or cigarettes in closets; adults may use matches as a light source in a dark closet or attic. Occasionally a lighted cigarette that is flipped out of a moving vehicle may be blown back in, unnoticed, and ignite the seats or floor mats.

Of course, the arsonist knows as well as fire investigators that these are obvious places where accidental ignition by open flame may occur. However, the arsonist cannot simply start a small fire, say in a wastebasket, and then hope for the worst. He must enhance the intensity and speed of travel to ensure destruction—perhaps with accelerants or by piling combustibles where they will surely become involved. And that is where fire investigators have the edge. They have the time, knowledge, and the tools to determine whether any aspect of a fire—not only the heat source—points to arson.

CASE HISTORY: Gas Range Fire

Firefighters responded to a fire in a large, well-known restaurant after the restaurant's normal closing time. The kitchen was well involved and firefighters noted black smoke as they advanced. At one point the fire was completely extinguished, but it reignited with explosive force. Since this was indicative of gaseous fuel, the main gas supply was shut off outside the building. The fire was then quickly extinguished.

Investigators found the point of origin to be in a large container of grease and other cooking waste. (Burning fats in the container had produced the black smoke.) The container was positioned directly on one of the burners of the restaurant range. The control for that burner was in the highest "on" position, and the control knob had been broken off.

No other ignition source could be found in the vicinity of the waste container. There was no sign of forced entry to the restaurant. These facts, the condition of the range, and the location of the waste container, pointed to arson, possibly by someone who had access to the restaurant. Questioning of the owner and employees confirmed that the restaurant had been losing money in spite of its excellent reputation.

The investigators called for an arson investigation and reported their findings as follows:

Point of origin: Grease container placed directly on burner of gas range
Heat source: Gas burner
Reason: Grease in waste container ignited by burner
Category: Arson.

5

FIREFIGHTERS' OBSERVATIONS

Chapters 2 through 4 were concerned with the investigator's *knowledge*—the things he needs to know in order to do an effective job. To a great extent, the remainder of this book is concerned with the investigator's *ability to observe,* which can be defined as the ability to see with the mind as well as the eyes.

We begin with a discussion of the process of observation as it applies to fireground investigation. Then, in the rest of this chapter, we discuss observations that may be made by investigators but are more usually supplied by firefighters (and sometimes by other witnesses). This information, like information gathered in any other way, should become a part of the fireground investigator's observations. The chapters that follow are concerned with observations the investigator himself is expected to make at the fireground.

THE PROCESS OF OBSERVING

Recall the case history dealing with solar ignition, in Chapter 4 (page 71). The first investigator examined the fire area carefully. At some time he must have seen the cut-glass decanter on the window sill above the point of origin, but he did not connect it with the fire. The second investigator saw the decanter, took note of its existence, and went a step further. He evaluated the decanter and its position in relation to (1) his knowledge of heat sources and (2) what he already knew about the fire. In other words, the second investigator *observed* the decanter.

We can use this example to define the process of observation, at least as it applies to fireground investigation:

1. *Sense* the object or condition. Usually, the eyes will be involved, but fireground sounds and smells can be important to an investigation.
2. *Notice* it. This means—essentially—allow the thing that was sensed to register in the mind. At first it may be necessary to force this step by actively thinking it, as, for example, "There's a cut-glass decanter in direct line with the sun and the point of origin." Eventually, this step should become second nature.
3. Briefly *evaluate* the sensed object or condition. Consider how it fits in (or doesn't fit in) with other facts concerning the fire, or with normal fire behavior.

This procedure will not magically transform everything that is seen, heard, or smelled into clues as to the fire's origin. However, it will help ensure that important evidence is not overlooked. As an example, consider the solar ignition case history. The second investigator might also have applied the observation procedure to the soap-powder box and the washing machine. (Both could have been involved in the ignition of the fire.) However, the third step—evaluation in light of other factors—led to their rejection as the heat source in that particular case; they had nothing to do with the fire. Careful observation did, though, turn up the actual heat source.

Evaluation might be thought of as a separate process, and not part of observing. The investigator will, after all, be making written notes about what he finds; can't he evaluate his findings at some later time? The answer is yes; the investigator will be continually considering what he found, until he finally determines the category of the fire. However, the act of noticing an object or condition should immediately and automatically cause the investigator to begin considering that object or condition in relation to the fire. The investigator cannot open his mind for the purpose of noticing an object, but then immediately close his mind so as not to consider that object. What we are discussing is a brief initial evaluation that should follow naturally from the act of noticing. This initial evaluation can be of help in eliminating certain possibilities (as with the soap powder and washing machine) or in directing the course of the investigation. If further evaluation of some object or condition then seems to be needed, the investigator may decide to put that off until some later time.

Perhaps the best way to settle this question of evaluation is to suggest that each investigator do what seems most comfortable. The investigator's findings must be evaluated at some point, and we are really discussing when this should be done. Often, the fire situation itself will dictate when certain findings should be evaluated. Then the key step in the observation process is to notice—to impress on the mind—what is seen, heard, and smelled. In fact, in the next several sections we discuss observations that may be made by firefighters and bystanders; in that case, we use the word "observe" to mean only to sense and to notice. Evaluation should be left to the investigator.

EARLY INFORMATION

Trained emergency personnel are expected to be capable of observing conditions and events at a fire scene; untrained persons may see and hear what is going on, but their recollections can easily be distorted by the inherent danger, excitement, and (for them) uniqueness of the situation. By trained personnel we mean fire investigators, firefighters, and police; untrained persons include occupants of the fire building, neighbors, passers-by, and other witnesses or spectators. And here we use the term "observe" to mean only "sense and notice."

Most fires (including arson fires) start unnoticed and are later discovered by untrained persons; some fires are accidentally started by untrained persons. But a trained observer is rarely on hand when a fire starts or is discovered. (The exception is the police patrol that discovers a late-night fire.) Yet the early stages of a fire, if observed, can yield information that is of great help to the fireground investigator. The investigator can gather this information only by questioning the somewhat unreliable, untrained witnesses. (Questioning procedures are discussed in Chapter 9.)

Once the alarm is turned in, the situation changes. Fire companies and police (that is, trained observers) are dispatched to the scene. The fire has made additional headway, but important aspects may still be observed. If the fireground investigator is among the earliest arrivals, he may observe the fire before and during attack. Most likely, though, he will arrive after firefighting operations have been started, and perhaps after extinguishment. But then he can rely on his questioning of police and firefighters for observations concerning the fire's earlier stages.

There are, however, two problems. First, although police personnel are trained observers, they are not usually trained to observe the details of a fire situation. In other words, they probably don't know what to look for. Moreover, police rarely enter the fire structure, so their observations are limited to the exterior. Second, firefighters, who are trained to observe the fire situation, will observe it from a firefighting—rather than a fire investigation—viewpoint. This is as it should be; firefighters are at the scene to protect lives and property, and not to worry about the point of origin, heat source, and so on.

The end result is the loss of important investigative information—because the earliest witnesses are untrained, and because the trained observers are involved in their own duties. One way to eliminate this loss of information is to dispatch a fireground investigator to the scene at every alarm. This solution is, however, close to impossible for most departments. A much more reasonable solution is to enlist the aid of firefighters in observing the situation from arrival to extinguishment. Responding firefighters will usually be the first trained observers at the scene; they have the knowledge needed to make the necessary observations; they will enter the fire structure to attack the fire; and they will be carefully observing the fire in performing their own duties.

Thus, if firefighters know what information is important to the investigation, they can usually observe and then supply that information. The additional concentration required of firefighters should not add to the burden of firefighting duties. Instead, it should increase firefighting efficiency—simply because crewmen will be more aware of the situation. Fire officers and fire investigators must ensure that firefighters understand what observations are needed and the importance of those observations. One way to do that is to discuss the material in this chapter with every fire crew in the district.

OBSERVATIONS DURING RESPONSE

Most responses to an alarm of fire are routine—not dull, but usually routine. On occasion, responding personnel may observe an unusual occurrence that will label a particular fire as being of suspicious origin. Anything out of the ordinary should be observed and reported; of particular interest are people hurrying off in the wrong direction and circumstances that cause a delay in the response.

Fleeing Persons

People are generally drawn to a fire; they congregate in groups to watch the fire and the firefighting operations. People in cars tend to slow down to get a better look, and some may stop. Thus, it is unusual for an innocent person to move rapidly away from the area. An individual who is observed leaving the scene, especially during the early stages of a fire, should be suspect. He or she may very well be an arsonist who has started an unplanned fire—perhaps in anger—or one whose planned fire has ignited prematurely, allowing little time for escape.

Responding personnel who notice a vehicle moving rapidly away from the scene should attempt to observe its color, make, model, license number, and any other identifying details, such as body damage. A person hurrying away from the scene on foot may be identified by approximate height, weight, type and color of clothing, and perhaps facial features (Figure 5-1). This information should be provided to the fireground investigator when the situation permits. Even if no identifying features are observed, the simple fact that someone did hurry away from the scene should be reported.

Forced Delays in Responding

The longer it takes firefighting units to respond to a fire, the more damage the fire will do. Knowing this, arsonists may attempt to delay the arrival of fire apparatus at the scene. One way to do so is to block the route to the

Figure 5-1. *The person on the sidewalk is fleeing the scene of the fire.*

fire—say, with a stalled vehicle on a narrow road or street, a tree felled or placed across a rural road, or even with stolen road department barricades of the type used to close streets to traffic.

Another delaying tactic is the false alarm that sends the apparatus in the direction opposite the arsonist's target. The false alarm, preceding the real alarm, lengthens the response time and allows the arson fire to gain additional headway.

Route blocking and false alarms are overt acts that should alert fire officers and investigators to the possibility of arson. More subtle delaying tactics are those that take advantage of an emergency situation or a condition that is not directly connected to the fire. For example, an arsonist may set a fire during a blackout or a heavy storm. Or the arsonist may time the fire to coincide with repairs on a road or bridge that is on the response route from the fire station to the fire structure. Anything that lengthens the response time is of help to the arsonist and so should be of concern to the fireground investigator.

CASE HISTORY: Delayed Response

At 6:10 p.m. an engine company was dispatched in response to the report of a bomb in a building housing a local radio station. The building was approximately southeast of the fire station. At 6:18 p.m., a fire call was received; a warehouse north of the fire station was burning. Ordinarily, the engine company was first due at the warehouse, but the bomb scare—false, as it turned out—delayed its arrival.

Other responding companies found one corner of the warehouse fully involved. The fire did not spread to the remainder of the building because it was held in check by a fully operating automatic sprinkler system.

The fire was extinguished, and an investigation begun. The investigator found enough evidence to categorize the fire as arson, several times over:

1. The false bomb alarm delayed the arrival of the first fire companies by several minutes.
2. The fire started in the only area that was not protected by an operating sprinkler system; the sprinkler section covering that area had been shut down for repairs several hours before the fire.
3. Flammable-liquid residue was found on packing materials in the area of origin.
4. There was no heat source within 25 feet (76 decimeters) of the area of origin.

The fireground investigator reported the following:

Point of origin: Packing materials near northeast corner of warehouse
Heat source: Probably a match or cigarette lighter
Reason: Deliberate ignition, using flammable liquid accelerant
Category: Arson.

The arson investigation team reasoned that only a warehouse employee would know that the sprinker section had been shut down. The investigation led to a worker who had been severely reprimanded and then fired by his supervisor. He admitted setting the fire for revenge, and using the bomb scare to delay the fire department response.

OBSERVATIONS ON ARRIVAL (BUILDING EXTERIOR)

Firefighters and/or investigators should observe conditions and events in and around the fire structure from the moment the first personnel arrive at the scene. Company officers must observe and assess the situation during size-up; at the same time, crewmen moving to assigned positions may observe elements of the fire that are initially hidden from officers' view. All such observations may be important to the fire investigation. Moreover, as fireground conditions change, the sequence of events should be observed and remembered. For example, there is a vast difference between

1. A slow, licking flame showing at one third-floor window, followed by sudden extension of flames out of several windows on that floor; and
2. Intense flames showing through several windows at an upper floor, followed by slow withdrawal of flames away from the windows.

In the first case, the change indicates the involvement of a fuel with the burning characteristics of flammable liquids. In the second case, the change indicates a transition from stage two burning to stage three burning (see Chapter 2).

Observations that may be made from the exterior of the building by arriving personnel fall into two groups: (1) the characteristics of visible flames and smoke and (2) those of the fire structure itself.

Flames and Smoke

Flame Location. Observe, as accurately as possible, each location at which flames are visible. Note whether the flames are confined within the structure or extend out windows, through the roof or roof openings, or through doorways. Note how much of the structure is involved, and the location of heaviest involvement.

If flames are visible at several locations within the structure, note how these locations are related to each other. For example, the flame locations may be above each other on the same side of the structure. Or they may be on the same level, at opposite sides of the building. (Widely separated flame locations may or may not indicate arson; however, this initial information will be of help in the investigator's later examination of the interior of the structure and his determination of the point or points of origin.)

Flame Color. The colors of visible flames can provide clues as to the nature of the burning materials and the intensity of the fire. Table 5-1 shows the flame colors produced by various materials that may be involved in structure fires. Note that the colors are not unique; that is, several substances may produce the same flame colors when they burn. Nevertheless, when considered along with such factors as normal building contents, burn patterns, and types of residues or debris, the flame colors can be used to determine what was burning.

Table 5-1. Flame colors produced by various materials

Color	*Meaning*
Yellow	Ordinary combustible. Class "A" materials such as cloth, wood, and paper.
Orange	Ordinary combustibles in latter stages of combustion.
Red	Flammable liquids, combustible liquids, and hydrocarbon by-products.
White	Metals such as magnesium.
Green	Copper and nitrates.
Blue	Alcohol and natural gas with a proper mixture of air.

Table 5-2 shows how the flame color changes as the heat production and intensity of the fire increase. The color makes a transition from light red (coolest flames) to orange to yellow and finally to white at the hottest point. Actually, this transition is one of brightness rather than color as in Table 5-1: The brighter and whiter the flames, the hotter and more intense the fire is. With a little experience, firefighters and fireground investigators should easily be able to distinguish between the flame color due to the particular material that is burning, and the flame brightness due to the intensity of the fire.

Table 5-2. Flame intensity (temperature) as indicated by color or brightness*

Flame color	*Temperature*	
Faint red	900°F	480°C
Red (visible in daylight)	975	525
Blood red	1050	565
Dark cherry red	1175	635
Medium cherry red	1250	675
Cherry red	1365	740
Bright red	1550	845
Salmon red	1650	900
Orange	1725	940
Lemon	1825	995
Light yellow	1975	1080
White	2200	1205
Blue white	2550	1400

*(from *A Pocket Guide to Arson Investigation,* 2nd ed. Boston, Factory Mutual Engineering Corp., 1979.)

Observe the color and brightness of the flames at each location where flames are visible. In particular, note the presence of red flames (indicative of burning hydrocarbons) at locations below the roof level. Note also the presence of unusual flame colors other than the orange and yellow flames produced by ordinary combustibles.

Flame Velocity. Along with the brightness described above, observations of the flame velocity can indicate the types of fuel and the possible use of accelerants. Observe visible flames, especially at open windows and doors and roof openings. Note whether they seem to be sluggish and without force (lazy floating flames, indicative of insufficient oxygen or minimal draft) or show considerable movement, reaching far out of openings in the structure (pushing or roaring flames, indicating strong draft and rapid burning). In particular, make note of unusually rapid flame spread, beyond the normal speed for the type of structure involved.

Smoke. Like flame colors, the color of the smoke is of help in determining what is burning (see Tables 5-3 and 5-4). In general, though, the earlier the smoke color is observed, the more meaning it has. As the fire progresses, fuels of different types become involved; then, rather than individual and distinct smoke colors, only a vague, dark mixture of the various colors may be observed. In addition, at the outset of a fire, the presence of thick, black smoke (which indicates burning hydrocarbons) could mean that flammable or combustible liquids were ignited. In later stages, the black smoke might indicate only that the roofing materials had become involved.

Table 5-3. Smoke colors and their meanings

Color	*Meaning*
Gray-white	Ordinary combustibles in early stages of fire
Dark gray	Ordinary combustibles in later stages
Black	Hydrocarbons; abnormal in earlier stages unless building contents include substantial amounts of hydrocarbon-based materials
Yellow-gray and brown gray*	Deep-seated, slow-burning fire; generally accompanied by heavy smoke stains on windows and little or no smoke movement*

*These characteristics also indicate the existence of backdraft conditions.

The density of the smoke produced by a fire can indicate the extent of burning, the presence of hydrocarbons, and as noted in Table 5-3, the existence of backdraft conditions. If an automatic sprinkler system is operating, the smoke may be mixed with steam. This can alter some visual aspects of the smoke.

Observe the color and density of the smoke and the location of the heaviest concentrations. If smoke is the only visible sign of fire, observe, in addition, the velocity of the smoke wherever it is leaving the structure and, if possible, how fast it is moving within the structure.

Windows and Exterior Doors

Windows and exterior doors may be opened, broken, or forced:

- By occupants fleeing the fire
- By persons making unauthorized entry into the fire building
- To increase the draft across the fire and, thus, the intensity and movement of the fire
- In the course of normal firefighting operations.

In addition, the heat of the fire may melt, crack, or break window or door glass.

After careful examination of the fire structure (Chapters 6 and 7) and questioning of occupants (Chapter 9), fireground investigators can usually determine how and why a window or exterior door was opened, forced, or broken. However, they need to know the condition of the doors and windows

Table 5-4. Smoke colors produced by various combustible substances*

Combustible substance	*Color of smoke*
Hay, vegetable compounds	White
Phosphorus	White
Benzine	White to gray
Nitrocellulose	Yellow to brownish yellow
Sulfur	Yellow to brownish yellow
Sulfuric acid, nitric acid, hydrochloric acid	Yellow to brownish yellow
Gunpowder	Yellow to brownish yellow
Chlorine gas	Greenish yellow
Wood	Gray to brown
Paper	Gray to brown
Cloth	Gray to brown
Iodine	Violet
Cooking oil	Brown
Naphtha	Brown to black
Lacquer thinner	Brownish black
Turpentine	Black to brown
Acetone	Black
Kerosene	Black
Gasoline	Black
Lubricating oil	Black
Rubber	Black
Tar	Black
Coal	Black
Foamed plastics	Black

*(from *A Pocket Guide to Arson Investigation,* 2nd ed. Boston, Factory Mutual Engineering Corp., 1979.)

when emergency personnel first arrived. They must also know whether any doors or windows were forced by firefighters in entering the structure.

Exterior Doors. Observe the condition or position of each exterior door prior to entry into the building by firefighters. The door positions may be categorized as:

- Open
- Closed but not locked
- Closed and locked (with bolt or chain)
- Blocked in the closed or open position
- Forced (but not by fire crews).

In addition, observe which exterior doors (if any) were forced by firefighters and what tools were used for forcing entry. (Burglar tools leave distinctive marks that differ in size and shape from the marks left by fire-service forcible-entry tools. However, it is not difficult to obtain fire-service type tools, so the latter information can be important.) Generally, firefighters should be able to recall where, when, and how they entered the building.

Windows. Observe whether any windows are open or broken prior to entry and venting operations (and, if so, which windows). In particular, note whether windows are open on both the leeward and windward sides of the building, to produce a natural draft.

Observe whether any windows are painted over or covered, or whether window shades are pulled down at an inappropriate time of day. Such techniques have been used (successfully) to hide fires from view during the early stages, and allow a longer burning time before discovery (Figure 5-2).

Observe which windows are opened normally or knocked out by firefighters for entry or ventilation. Note whether this work is performed from outside the structure or from inside. Normally, window glass falls inside the building when the window is broken from outside, and vice versa (Figures 5-3 and 5-4). However, each additional observation helps to support other observations—to produce a complete picture of the events prior to and during firefighting operations.

CASE HISTORY: Separate Fires

Firefighters responded to an afternoon alarm of fire at a six-story, multiple-unit dwelling. Few residents were at home. They found and extinguished fire in three separate locations—third floor front, third floor rear, and sixth

Figure 5-2. *Panel blocking window prevents early visual detection of the fire.*

Figure 5-3. A window broken from the outside.

Figure 5-4. A window broken from the inside.

floor rear. The fireground investigator, who arrived during extinguishment, was provided with the following early observations:

Observation	*Evaluation*
• Fire at front and rear of third floor—a separation of about 60 feet	• Separate fires could indicate arson

• Fire at rear of sixth floor, with no sign of fire at fourth and fifth floors	• Sixth-floor fire could have been set—but check for natural extension
• Gray smoke mixed with black smoke, red and orange-yellow flames at third floor front	• Hydrocarbons and ordinary combustibles involved
• Exterior doors closed but not locked; no obvious signs of forced entry	• Unlocked doors could allow unauthorized entry—check doors of involved units
• Many open windows in all apartments	• Accounted for by hot summer day

These early observations indicated that the fire might be arson.

Investigation of the interior of the building proved that there was a point of origin in a sofa in the third floor front apartment. The sofa was constructed with foam rubber padding and plastic upholstery, which could account for the black smoke and red flames. Burn and smoke patterns established that the fire had extended through the front apartment, out the open apartment door, and along a short hallway to the third-floor rear apartment. When that apartment became well involved, hot gases extended up a pipe chase, mushroomed into the sixth floor rear apartment, and flashed over. The result was a fire separated from the point of origin by two floors and the length of the building. The investigator ruled as follows:

Point of origin: Living room sofa in third floor front apartment

Heat source: Short circuit in table lamp cord

Reason: Cord was pinched behind sofa, shorted, and ignited upholstery on sofa back

Category: Accidental.

OBSERVATIONS DURING FIREFIGHTING OPERATIONS (BUILDING INTERIOR)

A fireground investigator may occasionally be on hand to enter and observe the fire building with attacking firefighters. However, the investigator cannot be everywhere at once, and most of the time he will not have arrived when the building is entered. Thus, in almost every instance the only available observations of the interior before and during firefighting operations will be those made by firefighters. For example, each fire department has its own procedure or pattern for making a complete and efficient search of the fire structure. During this search, firefighters are able to make observations that may be crucial to the fireground investigation.

Every observation discussed here need not be important at every fire; but each of them can be a key factor in a particular investigation.

Odors

The distinctive odors of flammable or combustible liquids such as gasoline, kerosene, turpentine, lighter fluid, and naphtha are most readily detected during the early stages of a fire. As the fire progresses, these liquid hydrocarbons are consumed; their residues produce practically no odor. Thus, on entering the fire structure, firefighters should use their sense of smell to observe any unusual odors—liquid flammables or others. These and other observations made within the fire building should be reported to investigators or fireground officers.

Sounds

Information concerning explosions that occur prior to firefighters' arrival at the scene may be provided by the dispatcher or witnesses. In addition, the fire structure may show evidence of a destructive explosion. During their attack on the fire, however, firefighters often hear the sounds of small explosions. In a dwelling, for example, the heat of the fire may cause an aerosol can or a paint can to explode. In a commercial or industrial building, supplies, solvents, or chemicals in sealed containers may explode. These "normal" sounds are indistinguishable from the sound of, say, the explosion of a container of flammable liquid placed in the building by an arsonist to produce a delayed acceleration of the fire. All should be observed and reported.

The sound of the fire itself can become so familiar that firefighters fail to observe it—it's simply there. However, a change in that sound, indicating an increase in the intensity of the fire, can be important to the investigation. The change may occur when a carelessly stored container of flammable liquid is knocked over by a hose stream. Or it could mean that a plastic bag full of gasoline was left in the path of the fire by an arsonist. Such changes in fire sounds should be observed and reported along with any other unusual sounds observed during firefighting operations.

Windows

Where feasible, windows should be observed from inside the building as well as from outside. Note whether the windows are open or closed or show signs of tampering. Observe, before opening any windows for venting, whether or not they are locked.

Interior Doors

Interior doors that are closed tightly will slow or stop the spread of fire. The arsonist will, therefore, do his best to ensure that interior doors are (and remain) open. He may wedge or block doors, especially fire doors, in the open position. One inconspicuous way to keep a door from closing completely is to place nails between the door and the jamb.

Firefighters should observe the positions of all interior doors. If a door is open or partially open, note whether it is blocked or wedged in that position. Note whether automatic fire doors have closed completely. If an automatic fire door should have been closed by the heat of the fire but was not, do not attempt to close it; do not remove any device that may have been used to keep it open. Do, however, report such abnormal conditions to the officer in charge.

The Fire

Observation from outside the fire structure can yield important information concerning the fire, especially in the earlier stages. However, what is observed from outside may not always indicate clearly what is occurring inside. For example, flames near the seat of the fire may be consuming hydrocarbons, but flames closer to windows may involve only ordinary combustibles; then the red flames typical of flammable liquids would not be observed from outside. Or, what might look like separate fires from the exterior could actually be the result of fire extension through an interior hallway.

For this reason, crewmen should observe the flames and smoke from inside the structure as well as from outside. In addition, some observations concerning the fire can be made only from within the fire building.

Seat of the Fire. Observe, as closely as possible, the location of the main body of fire. If possible, pinpoint the location to, say, a specific item of furniture in a dwelling, or a particular piece of machinery in a factory. If the room or area is heavily involved, it may be possible to observe only the approximate location of the seat of the fire, as, for example, in the storage area behind the sales floor.

Flames and Smoke. Observe the color and brightness of the flames. Note their velocity and intensity, and, if possible, determine whether the fire has vented itself. (A fire that is vented will show greater velocity than a fire that does not have a path out of the building; however, accelerants will add to the velocity of an unvented fire.) Observe the color, density, and velocity of movement of the smoke.

Paths of Fire Travel. Observe all paths of fire travel away from the seat of the fire and from any secondary fires. In particular, note any fire travel along a path or in a direction that does not seem to fit with normal fire behavior.

As an example, consider Figures 5-5 and 5-6. In both cases, the fire could be described as extending out of a room, along a hallway, and then up a stairwell. In Figure 5-5, the flames travel along the hall ceiling from the room to the stairwell; this is normal fire behavior. In Figure 5-6, however, the flames extend along the hallway ceiling *and floor.* This is abnormal for

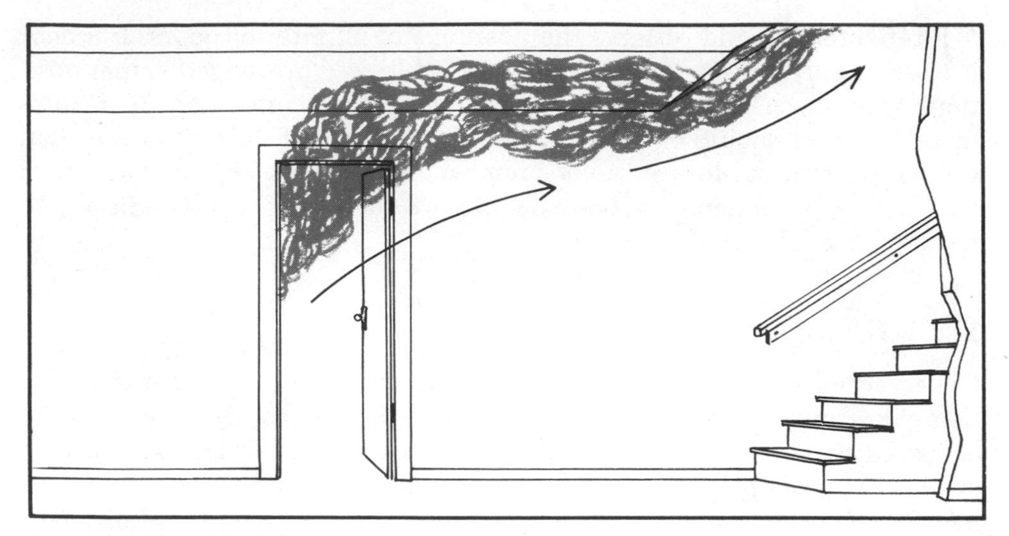

Figure 5-5. Normal fire behavior.

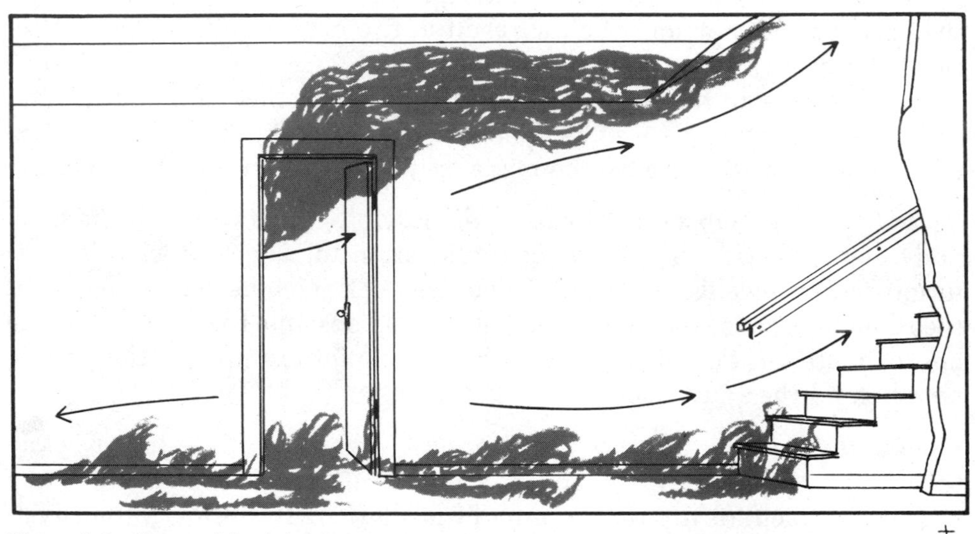

Figure 5-6. Abnormal fire behavior.

the conditions shown in the figure. Only a highly combustible fuel such as a flammable or combustible liquid placed on poured along the floor would lead to this erratic fire behavior. Note, also, the flames traveling toward the left of the doorway in Figure 5-6. Such flame movement, counter to the main flow of fire, would again be abnormal; it, too, indicates the presence of flammable liquids.

Arsonists sometimes make use of *trailers* to spread fire quickly through a structure. A trailer is a line of combustibles or accelerants (or a combination of the two) leading from the point or area of origin to some other area in the building (Figure 5-7). Although trailers are often quickly consumed

Figure 5-7. *Trailers accelerate the spread of fire throughout a structure.*

by the fire, they leave debris or fire patterns that are easily recognized. The discovery of a trailer is obviously very important in any fireground investigation.

Reaction to Hose Streams. When a straight stream is applied to fire in ordinary combustibles, flame production is usually reduced immediately. However, when a straight stream is applied to flammable liquid, the fire spreads in several directions and the flame velocity increases. This abnormal reaction is the result of the flammable liquid being pushed and splattered by the stream.

Similarly, a properly applied fog stream will quickly knock down fire involving ordinary combustibles. The fog droplets remove heat from the fuel on contact, and steam is produced; the fuel is cooled sufficiently so that the fire seldom will reflash. Flammable liquids and gaseous fuels have lower ignition temperatures than other materials, however, and will often reflash when attacked with a fog stream.

Firefighters manning attack lines can, therefore, provide important clues concerning the presence of flammable liquids. They should observe the reaction of the fire to hose streams—straight or fog—and report any abnormal behavior. They should also report:

- Any violent reactions that occurred when water was directed onto the fire (for example, magnesium may explode on contact with water, spreading fire in all directions)
- Agents other than water that were needed to achieve extinguishment, and the amounts required
- Situations in which an abnormally large amount of water was required for extinguishment, given the specific type of fire and the fire load.

CASE HISTORY: Too Much of a Bad Thing

At 2:40 a.m. in a fairly large city, fire companies responded to a fire in a clothing store. The store was on the street floor of a two-story building; the second floor was unoccupied and was used mainly for storage. First-due companies found substantial fire just inside the store's display windows; red flames reaching out through the open front windows of the second floor; and black smoke, but no flames, venting through the roof.

The fireground investigator, who had responded with first-due companies, made or was provided with the following observations:

Observation	*Evaluation*
• Fires on both the first and second floors	• Possibly separately set
• Red flames and black smoke	• Hydrocarbons burning—unusual for a clothing store
• Smoke issuing from roof	• Unusually fast burn through at roof
• Front and rear doors had to be forced by firefighters	• Normal for a store in early morning
• Attack crews report strong odor of gasoline	• Abnormal in a store—used as accelerant?

• Extinguishment difficult; flames accelerate and spread when streams are applied	• Flammable liquids involved—probably the gasoline
• Fire on stairs to second floor; full involvement, difficult to control	• Abnormal for fire to burn low—accelerant used on stairs?
• Ladder truck crew reports cover removed from roof scuttle	• Explains venting of smoke through roof; also accelerated vertical spread

At this point, even before entering the building, the investigator designated the fire as arson. The fire marshal was called in, and the fireground was declared a crime scene.

Examination of the premises revealed the work of an overly thorough arsonist. In the store itself, a number of small rectangular cans, containing gasoline, had been placed close enough to each other to create a "domino" effect. When a hose stream knocked one can down, that can knocked the next can down, and so on, causing the fire to spread away from the entrance. A char pattern, in an almost perfect straight line from the middle of the wooden floor to the stairway, was obvious evidence of the use of a trailer. On the stairway itself, plastic bags filled with gasoline had been placed on every other tread. As the bags were melted by the heat, the gasoline carried the fire up to the second floor. For some reason, the heat and flames had not affected two of these devices; they were found completely intact.

Fire marshals and detectives also uncovered evidence of the poor financial status of the clothing store. The owner was eventually tried and convicted of arson; he then implicated the overly enthusiastic arsonist he had hired.

Obstructions That Slow Firefighting Operations

To gain additional burning time, an arsonist may set up obstructions that will block or impede firefighters' access to the fire. The arsonist may attempt to block entry to the fire structure or to particular areas within the structure; the obstructions may be placed outside the building or within it.

For example, access to manufacturing facilities or warehouses can be blocked by positioning trucks or stacks of heavy merchandise at loading-platform entryways; positioning fork lift trucks or heavy crates at the entrances to storage or fabrication areas; and boarding up windows or piling contents against windows. Most of these techniques reduce the effect of venting operations and prevent hose streams, as well as firefighters, from reaching the fire. In a dwelling, furnishing may be piled against windows or interior or exterior doors, to achieve the same results.

The arsonist's use of a *drop lock* can transform a door with a flimsy lock into an obstruction that must be battered down for entry. The drop lock is a heavy board with a string attached to it. The board, on the inside, is maneuvered into a wedged position against the door by means of the string, which passes over the slightly open door to the outside. When the board is wedged, the door is closed completely. The fire consumes the string, leaving no trace. When the door is finally opened, the board is knocked aside to become just another piece of debris.

A particularly malicious method of impeding access to the fire is to remove a section of flooring, just inside a doorway. Arsonists have also removed stairways to keep firefighters from reaching upper floors of the fire building. These obstructions to firefighting activities—and any others, no matter how "natural" they may seem—should be reported to fire officers and investigators.

Building Contents

Fireground investigators are interested in any departure from the normal contents of the fire building—in amount, type, or position. However, firefighters cannot usually be expected to notice such differences unless the contents are quite obviously wrong for the occupancy. The one exception is in business establishments that firefighters have visited earlier, during fire-prevention inspections. There, they should be able to notice any changes or reductions in the contents between the time of the inspection and the time of the fire.

Firefighters should also observe and report instances in which:

- The contents of a dwelling or business are obviously meager, or some contents are obviously missing
- Contents are scattered about, with drawers opened or removed, in a disarray that indicates vandalism or burglary
- Furniture is piled together either near the seat of the fire, to burn more readily, or near points of entry, to block access (Figure 5-8)
- Crates, cartons, merchandise, or other items are stacked so as to block the flow of water from a sprinkler head (Figure 5-9) or to block stairways, automatic doors, or points of access
- Crates or cartons are broken open to expose combustible materials to the fire.

Building contents are discussed from the investigative viewpoint in Chapter 7.

Fire Protection Systems

If a target structure is protected by an automatic fire alarm and/or extinguishing system, the arsonist will attempt to keep the system from operating properly. Firefighters should know, through preplanning and fire-prevention inspections, the type of system installed and how it is supposed to operate. They should observe and report any abnormal operation.

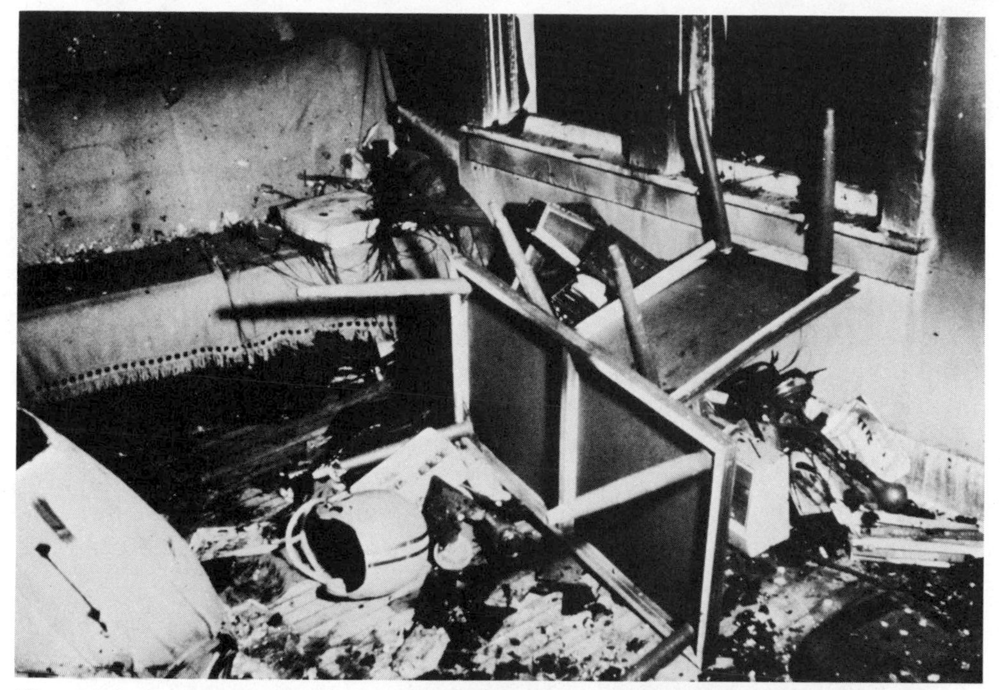

Figure 5-8. *Check unusual positioning of room contents for possibility of crime cover-up.*

Figure 5-9. *The pallet had been placed above the burning material to shed water.*

An automatic sprinkler system should detect and confine fire with a maximum of five sprinkler heads operating. Each head should provide a uniform spray pattern, 8 to 10 feet (24 to 30 decimeters) in diameter. There should be sufficient pressure to maintain that pattern. Firefighters should report the failure of sprinkler heads to operate, weak sprinkler-head patterns, or any fire that activates more than four sprinkler heads. (Fire departments pump into sprinkler systems via siamese connections that bypass the main control valve. Firefighters may therefore see adequate spray patterns during pumping. The flow from the heads should be observed after the external water supply is shut down.)

The failure of more complex automatic extinguishing systems, such as Halon and CO_2 total flooding systems, may be difficult to observe. The gases used to flood the protected space are invisible, and so their absence would not be noticed by firefighters using breathing apparatus. However, firefighters can observe and report the fact that such a system did not extinguish the fire. Moreover, an arsonist can destroy the integrity of the protected space by blocking its door in the open position; this allows the extinguishing agent to escape, so that the fire is not controlled. Again, this condition should be noted and reported.

In addition, investigators should be made aware of (or should question firefighters concerning):

- Standpipe systems that did not operate properly
- Automatic fire-detection devices that failed to transmit the alarm, or that sounded a delayed alarm
- Explosion suppression systems that failed to operate.

Firefighters' Actions

Along with everything else, firefighters must be able to describe, in detail, the actions they took on the fireground. We have already noted the importance of observing how entry was made and the condition of windows and doors prior to entry. Personnel manning hoses should be able to recall how the hose lines were stretched, the points from which streams were applied, and the stream patterns that were used. Ladder truck crews should remember where ladders were raised and where openings were made for venting or entry (as, for example, by forcing roof scuttles).

The time at which utilities (gas, electric, water supply) are turned off should be noted, along with any changes in the fire situation that result from this action.

If any of the building contents must be moved during firefighting operations, their original positions should be carefully noted. By moving a piece of machinery or an appliance, firefighters may be changing the location of the heat source; by moving burned contents, firefighters may disrupt the continuity of the burn patterns or remove the point of origin from contact with the heat source. It is extremely important that the investigator know of such changes in location.

Personnel engaged in overhaul and cleanup should be especially aware of this problem. They should note both the original location and the final disposition of any debris that was moved during these activities. Items that must be removed from the building should be carried, rather than thrown out of windows.

Evidence

Arson devices range from containers that may have held flammable liquids to sophisticated timing mechanisms. A firefighter who encounters such a device (or something that looks like it might be such a device) should, if at all possible, leave it in place. *By touching or moving evidence, firefighters can destroy its acceptability in court.* Burned materials near the arson device must also remain in place, so that arson investigators can find and prove the connection between the device and the first material to burn.

It is to the benefit of firefighters as well as investigators to protect evidence of arson. And the best way to protect evidence is to keep it intact. During attack and extinguishment, personnel will necessarily concentrate on their firefighting duties, rather than on the preservation of evidence. However, if possible evidence of arson is found during or after extinguishment, overhaul and cleanup can certainly be delayed until investigators have had a chance to assess the situation.

CASE HISTORY: A "Minor" Observation*

At 10:21 p.m. on a late summer evening, fire companies were dispatched to a fire in a single-family dwelling in a suburban area. The occupants were not at home; the alarm had been called in by a neighbor. Crew members made the following observations on arrival and during firefighting operations:

Observation	*Evaluation*
• Fire showing on first and second floors	• Possibly separate fires, but inconclusive alone
• Yellow-orange flames with some red flames; gray smoke intermingled with black	• Both ordinary combustibles and hydrocarbons involved
• Smell of gasoline in interior	• Accounts for red flame and black smoke—used as accelerant?
• Furniture in disarray; drawers pulled out of chests; bedroom closets and bureaus open, with clothing strewn about	• Possible burglary with arson as a cover-up

*This case history is adapted from "A Criminal Trial Lost—A Civil Trial Won," reported by D. B. Lemieux in *The Fire and Arson Investigator,* vol. 31, no. 1, p. 59, 1980.

• Living-room drapes pulled closed, but destroyed by fire; bedroom window shades down	• Not unusual for time of fire—but occupants were not at home to pull drapes and shades; possibly done to hide early stages of fire
• Empty 2-gallon gasoline can found in uninvolved cellar	• Possible source of accelerant—but why was it placed in the cellar?

The senior fireground officer declared the fireground a crime scene and called for an arson investigation. The owners of the dwelling, who were vacationing some distance away, were summoned to the scene the next day. They stated that several valuable items were missing from the dwelling, including sterling silver tableware, gold jewelry, two antique paintings, and an oriental rug. They also noted that the gasoline can was usually kept in the cellar, exactly where firefighters had found it.

The neighbor recalled hearing a vehicle in the driveway of the involved dwelling, an hour or so before discovering the fire; she did not, however, see the vehicle clearly. She did tell the investigator that the owners of the dwelling had been unsuccessfully trying to sell their house for several months, in order to move to another area.

The investigator was bothered by the location of the gasoline can. Laboratory analysis had determined that traces of gasoline in the can matched residues found in wood flooring on the first and second floors. The investigator reasoned that one or both owners could have returned to the house, pulled the shades and drapes, removed the valuables, thrown furniture and clothing about to give the appearance of a burglary, set fire to the dwelling, and then driven back to the vacation resort. He felt that whoever poured the gasoline in the house had returned the can to the cellar through force of habit; a burglar or arsonist would not have developed that habit. The investigator could find no trace of the missing valuables or of a burglar.

The case was brought to trial, but the owners were acquitted. The judge felt that there was a reasonable doubt that they had committed the crime. The insurance company, however, refused to pay off on the fire insurance policy, noting that there was some evidence that the owners had set the fire. The owners sued the insurance company in civil court. The evidence was almost exactly the same as in the criminal case, but here the owners *lost.* The civil judge felt that the owners had to prove that they didn't set the fire, and they were unable to do that.

Obviously, the firefighters' observation of the gasoline can in its usual place was very important. However, this case history has another important point: Even when the criminal trial of an alleged arsonist is lost, a civil action may remove the profit from arson. Cooperation among fire department, law enforcement, and insurance company personnel is essential.

OBSERVING SPECTATORS

Fire personnel assigned to duties outside the fire structure have the opportunity to observe spectators and, thereby, aid in the investigation. For example, once pump operators have supplied the attack lines, they can watch and listen to the comments of people witnessing the fire. Fireground investigators who arrive during firefighting operations should observe the spectators as well as the fire. Police officers at the scene can observe while performing other duties.

Many spectators seem almost compelled to make comments about the fire and occupants of the fire building. Some of these comments can contain valuable information, sometimes without the speaker's realizing that fact. A neighbor who comments to another that "They're lucky—they just put half their furniture in storage" should certainly be questioned by investigators. Or, an unsubtle remark like "This is the second time Joe's business has been saved by artificial lightning" may or may not contain some truth, but it should not be ignored—especially if other observations indicate the possibility of arson.

Emergency personnel should also be alert to spectators who volunteer their help in stretching hose lines, raising ladders, or rescuing occupants. Disturbed persons will set fires simply for the chance to help fight them; in some cases, such people have risked their own lives rescuing victims at fires they have set. Once the fire is under control and the excitement has diminished, these disturbed firesetters tend to melt into the crowd. Other firesetter types are those who seem to show up at every fire, and those who are particularly excited—sometimes even hypnotized—by the fire and the activity.

We have, in this chapter, discussed perhaps fifty ways in which working firefighters can contribute to fireground investigations. A firefighter who is unfamiliar with fireground investigation might respond, with some justification, that he would be perfectly willing to do all the observing that is needed, if only someone else would perform his fireground duties. The truth of the matter is, however, that firefighters can do—and are doing—both. Progressive fire departments across the country are now performing this function. In departments with newly organized fireground investigation functions, it is the job of officers and investigators to promote the concept and demonstrate that it is an important one.

6

INVESTIGATING THE FIRE: BUILDING EXTERIOR

The investigation of a building fire is comprised of three major parts: examination of the exterior of the building, examination of the interior, and questioning of firefighters and witnesses. Each part of the investigation should produce information that supports and extends the information gathered during the other two parts. Taken together, they should give the investigator an accurate view of the circumstances that led to the fire.

There are no set rules for determining how much time or effort should be devoted to each part of the investigation. Instead, the particular fire situation should indicate what is needed. A fire that remained small because it was discovered and extinguished quickly may not even require all three investigative phases. On the other hand, information that is learned, say, during questioning, may cause the investigator to perform a very careful, painstaking, and time-consuming examination of the interior of the structure.

Nor are there any rules as to which part of the investigation should come first. Ideally, the fire would be investigated by a team with at least three members, so that all three parts may be performed at once. However, this is seldom possible. Most investigators who work alone prefer to begin with the exterior examination and then move into the building, asking questions of firefighters and witnesses as opportunities allow. They complete the investigation by interviewing any witnesses who may have been missed earlier.

In keeping with that sequence, the exterior examination is discussed first, in this chapter. We begin, however, with several topics that concern all three phases of the fireground investigation.

TOOLS AND EQUIPMENT

A variety of tools, equipment, and supplies may be required for the examination of the fire scene and the collection and preservation of evidence. Listed below are the contents of a typical investigative equipment kit; individual investigators will, of course, modify these contents to suit their own experience and preferences.

1. *Protective clothing*
   Helmet
   Coat
   Gloves
   Boots
2. *Hydrocarbon tester*
3. *Measuring devices*
   Folding rule
   100-foot measuring tape
4. *Lights*
   Flashlight
   Heavy-duty light
5. *Cutting tools*
   Penknife
   Hand axe
   Handsaw
   Keyhole saw
   Hacksaw
   Cold chisel, wood chisel
   Scissors
   Pliers (wire cutters)
6. *Tools for collecting evidence*
   Shovel (small, flat end, with short handle)
   Tongs (kitchen type)
   Tweezers
   Magnet
7. *Prying tools*
   Pry bar
   Claw hammer
   Screwdriver
8. *Sifting tools*
   Claw tool (garden type)

Quarter-inch-mesh sieve, about 2 feet square
Strainer (kitchen type)
Whisk broom
String (to mark out search area)

9. *Photographic equipment*
   Camera
   Film
   Flash device
   Note: The type of photographic equipment should be compatible to the expertise of the user.
10. *Containers for evidence*
   Unlined cans (plain metal interior)
   Glass jars with caps
   Boxes (ranging from pillbox size to cartons)
   Flat pieces of glass
   Styrofoam cups
   Paper bags
11. *Sealing and marking materials*
   Masking tape
   Sealing wax
   Wrapping paper
   Tags (with strings or wire)
   Pen with waterproof ink
   Felt-tip markers (waterproof)
12. *Note-taking and diagramming supplies*
   Clipboard
   Quarter-inch grid paper
   Small straightedge
   Pencils with erasers
   Fine-tip colored marking pens (blue, green, red, orange, black)
   Notebooks
   Portable tape recorder, extra cassettes
13. *Materials for protecting evidence on the fireground*
   Rope or clothesline (for cordoning off an area)
   Warning pennants or signs ("Fire line—do not cross")
   Lightweight plastic tarpaulin
   Small traffic markers (various colors)
14. *Miscellaneous*
   Magnifying glass
   Cigarette lighter (for melting wax)
   Scribing tool (to mark evidence)

The use of most of this equipment will be discussed in this chapter and the next two chapters.

PHOTOGRAPHS

The camera is perhaps the most useful piece of equipment in the investigator's kit. Photographs can provide an accurate record of the extent of involvement, the extent of damage, the placement of evidence when found, the location and configuration of the low burn, and any other features that might have to be recalled exactly—either for analysis of the fire or as evidence in an arson trial.

Camera Use

Although some fire departments may be able to dispatch photographers to assist in fireground investigations, in most cases the investigator must take his own photographs. Even a novice can produce clear, useful photos if he uses good photographic equipment, adheres to the manufacturer's recommendations, and follows a few simple guidelines.

1. Take color prints, rather than slides. They are easier to refer to later.
2. Start taking photographs at the beginning of the fireground examination, and take them as needed throughout the entire examination. (Do not leave the photographing until after the scene has been investigated.)
3. Take more, rather than fewer, photographs. (It is not unusual to take 80 to 100 photos during a routine fireground investigation.) If in doubt as to whether or not to photograph an item of evidence, take the photo.
4. Include the structure's building number (address) in at least one photo.
5. Take a photograph of the nearest street sign, including the fire scene in the photo if possible.
6. Take two photographs of each piece of meaningful evidence, in the position in which it is found. First take a distant photo showing the evidence and its surroundings, for ease of identification and location. Then take a close-up photo, to allow detailed study of the evidence.
7. Establish the size of the evidence in each close-up photo by placing a ruler nearby and including it in the photo. If a ruler is not available, any common item (a cigarette package, for example) will serve to establish relative size.
8. Make sure the chain of custody is maintained for both undeveloped film and developed film and prints (see Chapters 8 and 10).

Identification of Photos

Each photo that is taken on the fireground should be identified in two ways. First, the investigator should record, in his notebook, what was photographed, its location, the time the photo was taken, the approximate distance from the camera to the subject, and the number he has assigned to the photo. (The numbers assigned to photos must correspond to the numbers assigned to items of evidence. A standard numbering system is discussed in Chapter 8.)

Second, when the film is developed and printed, each print should be marked with the information that identifies it. A rubber stamp, in the format shown in Figure 6-1, will remove most of the drudgery from this chore and serve as a reminder that it needs to be done. If an instant (self-developing) camera is used, each print should be marked as soon as it has developed.

FIRE DEPARTMENT

ALARM NO.______ LOCATION____________________

DATE OF PHOTO________ 19____ TIME OF DAY______

SUBJECT MATTER____________________

DISTANCE LENS TO SUBJECT______ FT.

DIRECTION OF PHOTO__________

PHOTO BY:__________

DEVELOPED BY:____________________

REMARKS____________________

Figure 6-1. Necessary information is stamped on back of all photos.

Marked photographs should be placed in an envelope for protection and stored with other records pertaining to the fire.

WRITTEN NOTES

During the examination of the fireground and questioning of witnesses, investigators gather a lot of information in a relatively short time. It is extremely difficult to remember all this information—to keep track of who said what and which piece of evidence was found where. For this reason, investigators must make notes concerning what they do and what they observe during each phase of the investigation.

The format of the notes and the type of notebook used are matters of personal preference. However, the notes for each investigation should begin on a clean page (or even in a new notebook) with the date, time of arrival, address of the fire scene, and incident or response number. In addition:

- Each pertinent observation should be described briefly, and the time and place of observation noted. The initial evaluation of the observation may also be included.
- The essence of each interview should be noted, along with the name of the witness (or firefighter) and the time of the interview. Here too, an initial evaluation of the information or the interviewee may be included.
- Information regarding each photograph taken at the scene should be placed in the notebook (as discussed in the previous section).
- When a diagram or sketch is drawn, a notation should be made as to what was sketched, the angle or direction of the view, the time the sketch was made, and the reason for making the sketch.
- If the point of origin, heat source, reason, and category are determined at the fireground, they should be listed in the notebook.

In spite of the rush of information, these notes should be legible and accurate, so that they can be re-read and typed up as part of the permanent record of the fire investigation. The notes, along with any photos or sketches made at the scene, are evidence that may be used in court—sometimes months or years after they were made.

Some investigators prefer to record their observations, interviews, and impressions on a portable tape recorder. However, they also find it necessary to make written notes concerning photos, sketches, and other aspects of the investigation.

DIAGRAMS AND SKETCHES

Diagrams and sketches should be used to show building features, the extent of the fire, smoke and flame patterns, locations where evidence was found, and other important information. In some situations, a single diagram may show the entire involved area; in others, the investigator may want to prepare one diagram for each room or level that was involved, and several diagrams showing the exterior of the fire structure. The diagrams and the photographs taken at the scene support each other; together, they provide strong evidence of the behavior of the fire if legal action is instituted. For this reason, it is important that there are no conflicts between the information contained on photos, on diagrams and sketches, and in the investigator's notes. All the information must be accurate and consistent.

Diagrams and sketches should be kept simple. It is easiest to draw them with a soft pencil on quarter-inch grid paper, held in place on a clipboard. The grid lines are of help in getting straight lines straight and in approximating distances. However, *diagrams and sketches must not be drawn to scale.* A scale drawing has to be *completely* accurate; a single inaccuracy will enable a lawyer to discredit the entire drawing. There is no such restriction on a drawing that is not to scale but is simply used for reference.

Notations that pertain to items shown on the diagram should be included—for example, "Top window glass broken outward" or "Gasoline can found here." Where a measurement is important to the investigation, it too may be noted on the diagram. The measurement should, of course, be accurate.

Figure 6-2 is an investigator's diagram of the situation shown in Figure 6-3; Figure 6-4 is a diagram of the wall in the photo of Figure 6-5. Note the features that are drawn and pointed out on the diagrams. This is what makes diagrams so important, even though they may show the same conditions as the photos: The investigator can emphasize features and make annotations on a sketch, but not on a photograph.

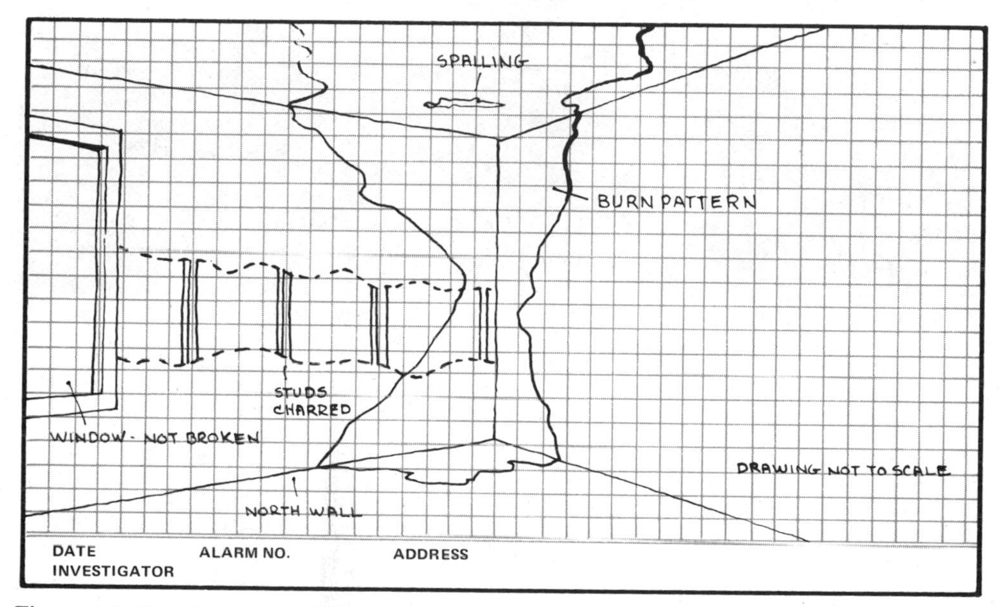

Figure 6-2. Investigator's sketch of burn pattern in Figure 6-3.

As with photographs, each diagram should be identified in the investigator's notes. For example, for Figures 6-4 and 6-5, the investigator might include, in his written or recorded notes, the identification, "Very obvious V pattern enclosing doorway in north living-room wall; see interior sketch 3, photo 8." The photo would be further identified, as discussed earlier.

EXTERIOR INVESTIGATION

The exterior investigation includes examination of the fire building from the outside and examination of the grounds surrounding the building. Each should be examined individually, in turn. That is, the investigator should

Figure 6-3. *The point of origin was in the corner with mushrooming at ceiling.*

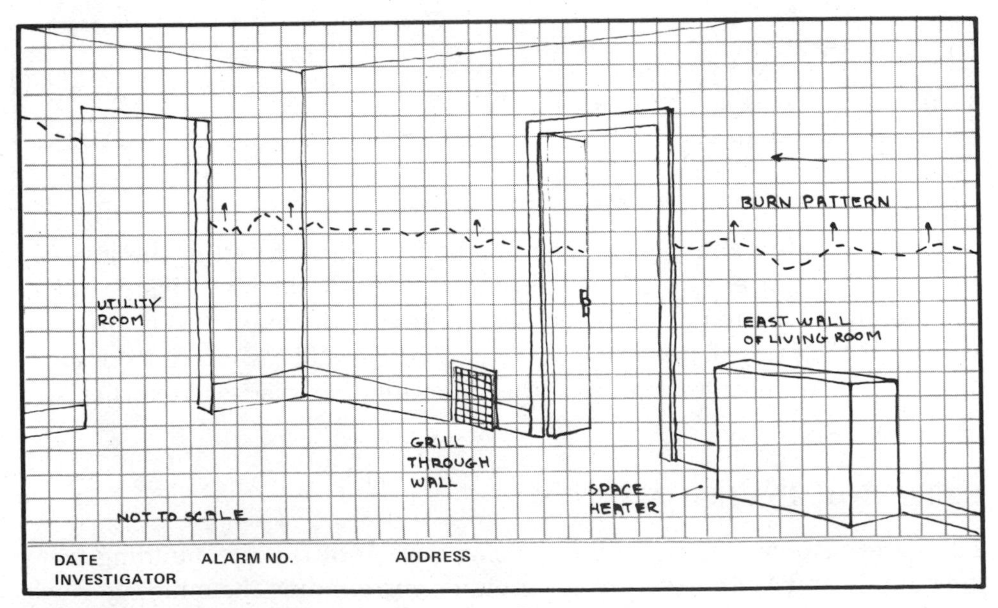

Figure 6-4. *Investigator's sketch of burn pattern in Figure 6-5.*

Figure 6-5. *Burn patterns move upward on the walls as they move away from the point of origin.*

walk around the building at least once to examine the grounds and at least one more time to examine the fire building. (If an exposure was involved, it too should be investigated to determine whether the fire spread to or originated in the exposure as a separate fire.)

Maintaining the Integrity of the Fireground

Unauthorized persons should be prevented from entering the fire structure and the surrounding area from the time firefighters first arrive until the investigation is completed. If necessary, the ground should be roped off and declared off limits to unauthorized personnel. Police, if available, should be requested to enforce this restriction.

A fireground that is not restricted in this way is open to theft and the accidental or purposeful removal and destruction of evidence. If occupants later report something missing, the theft might be assumed to have occurred before the fire, when it actually was the result of looting. Such theft might also be blamed on firefighters. Or an arsonist might have a chance to remove or destroy evidence that remained after extinguishment. On the other hand, evidence might accidentally be kicked, stepped on, moved, or otherwise disturbed by occupants anxious to survey the damage done by the fire.

Perhaps the best reason for keeping unauthorized persons away is the possibility that the fireground might have to be declared a crime scene. Then, if the integrity of the fireground has not been maintained, evidence of arson might not be credible in a court of law. A defense attorney could easily show that the fireground was contaminated (by the presence of unauthorized persons); the evidence would be considered contaminated—and therefore unreliable—as well.

At best, the fire investigation becomes more difficult when unauthorized persons are permitted on the fireground. At worst, the contamination of evidence might allow an arsonist to go free, or cause an innocent person to be suspected. If the senior fire officer has not already cordoned off the fireground, the investigator should do so immediately upon arrival.

Surveying the Scene

It is a good idea to take a long look at the entire fire scene before beginning the first tour around the building. Observe how the building is situated on the ground; the locations of driveways, power lines, and outbuildings; whether or not the grounds seem to have been involved in the fire; any trees or shrubbery that might have had an effect on the fire, its discovery, or an arsonist's tactics. In other words, get a feel for the entire situation. Then begin the first exterior inspection.

Walk around the outside of the structure slowly. Take time to *observe* as well as *sense* what is there. Record anything that appears to be connected with the fire; what may seem unimportant when first observed can prove to be most important when it is later analyzed. Treat all evidence as detailed in Chapter 8.

EXAMINING THE GROUNDS

The area surrounding the fire building may yield nothing, or it may produce a wealth of information. There is only one way to find out—and that is by performing a thorough examination.

Uninvolved Grounds

If only the building was involved (and not grass, brush, or trees), the examination of the grounds may be limited to a strip about 25 feet (76 decimeters) wide surrounding the fire building. (It is true that evidence of arson has, on occasion, been discovered hundreds of feet from the fire structure. However, if the fire is found to be of suspicious origin, the search for evidence can be expanded by investigators.) Of course, if something of importance is noticed beyond the 25-foot limit, it should be investigated.

During a slow tour of the fire building examine the ground for clues or objects related to the fire. Unfortunately, it is impossible to be specific about what to look for; that depends on the particular situation and usually isn't known until something is found. However, we can list some general categories.

Containers. Bottles, jars, cans, milk cartons, and plastic bags that are found near the fire building may be significant. These containers could have held a flammable liquid that was used to ignite or accelerate the fire. If so, they may give off the telltale odor of a liquid hydrocarbon. Even the pieces of a broken glass container can yield flammable-liquid residues in laboratory tests.

Tools. Hammers, pry bars, screwdrivers, tire irons, or other tools found on the fireground may have been used for illegal entry into the fire building. The discovery of such a tool should alert the investigator to the need for very careful examination of the building's doors and windows. It is often possible to match a prying tool with the pry marks produced by that tool.

Valuables. Such items as money, silverware, jewelry, furs, and stamps or coins from collections, discovered outside the fire building, could indicate that the fire building was burglarized. However, they could also have been dropped by occupants rushing to save their belongings. Examination of the building and questioning of occupants should clarify the situation.

Footprints and Tire Tracks. The footprints made by firefighters' footwear are distinctive and easily recognized. Prints left by other types of footwear—especially under windows—should be protected from destruction; they may have to be photographed and used as evidence if the fire is categorized as arson (Figure 6-6). Footprints may be protected with the evidence-protection materials listed earlier in this chapter or, if necessary, by personnel stationed nearby for this specific purpose.

Tire tread marks that lead to the area behind the fire building or to some other concealed area should be cause for suspicion. Such tracks may have

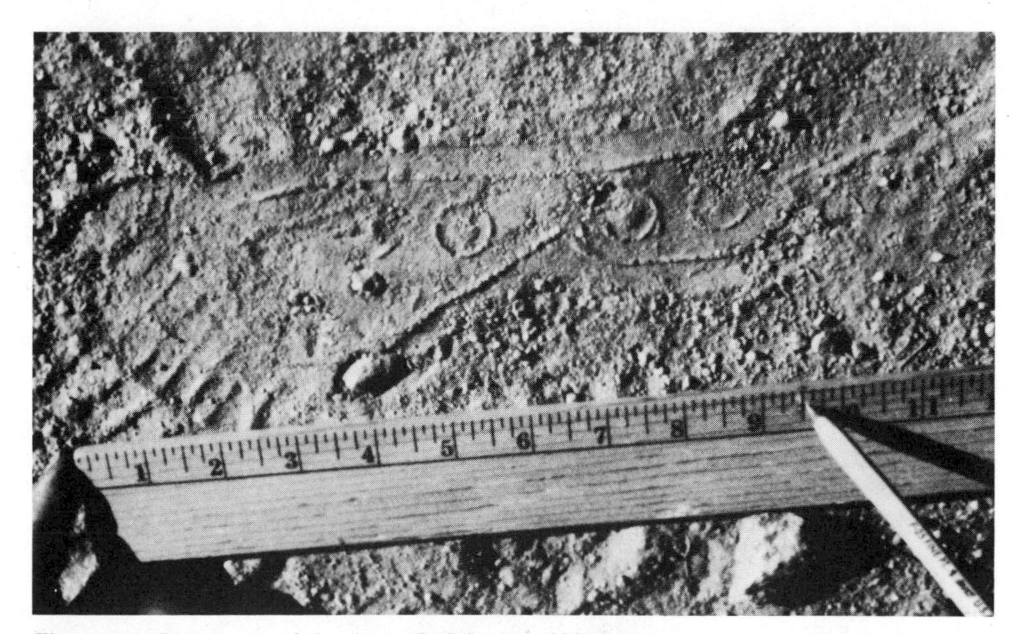

Figure 6-6. *Footprints not belonging to firefighters could be evidence.*

been made when a vehicle was parked out of sight. The tracks should be preserved until the fire is categorized (Figure 6-7).

Grounds Involved with Fire

When grass or brush is involved, the entire burned area of the grounds should be inspected, in addition to the area around the fire building. The investigator should, as before, look for any objects or clues that seem to be related to the fire. In addition, he must determine whether the fire originated outside and then traveled to and into the structure, or whether fire extended from the structure to the grounds. In the former case, the fire may have started in brush or grass, or in a pile of trash. Such a fire can travel some distance by burning live grass and dry grass clippings at ground level. On occasion, vehicle fires have extended to buildings through direct flame contact or radiant heat.

Grass Burn Patterns. Grass fires travel most rapidly with the wind. As they do, they leave burn patterns that narrow to a point. The flames will also move against the wind, but much more slowly; the burn pattern is then much wider, because the wind tends to cause the flames to fan out. Thus, if the wind was blowing toward the building from the ground fire area, and the grass burn pattern "points" to the building, the fire probably started outside.

Structure Burn Patterns. A grass fire that moves to a building leaves a V-shaped burn pattern on the outside of the building. The V can be traced

Figure 6-7. *Observe tire tracks from vehicles.*

upward to the point at which it terminates or to where the fire entered the building. Any windows within the V should be stained (by smoke) on the outside. Where the fire entered the building, the burn pattern should continue upward on inside walls.

Figure 6-8 is an investigator's diagram of a fire that originated outdoors in some trash and spread to a house. Note the point of the grass burn pattern, the V-shaped burn on the house wall facing the wind, and the several notations included on the diagram.

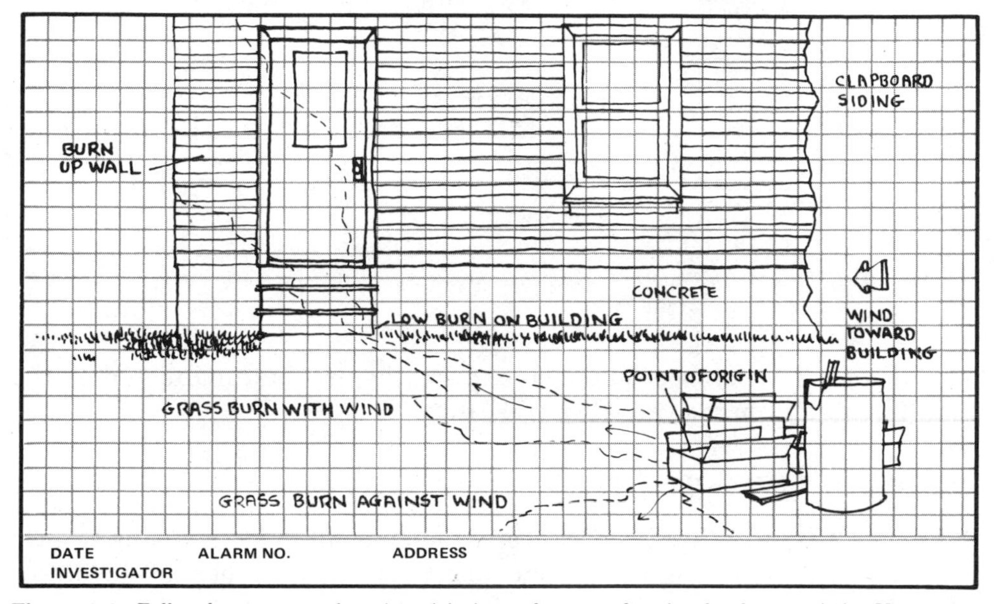

Figure 6-8. *Follow burn pattern from its origin in trash can and notice the characteristics V pattern.*

When fire spreads from within a structure to grass or brush, the method of fire travel is almost always embers dropping from the structure. Then, burn patterns within the structure will show fire travel toward the opening through which the embers fell. Smoke stains will be found on the inside surfaces of windows. And there will usually be no interconnection between the burn pattern on the outside of the structure and that on the grass.

Figure 6-9 is an investigator's diagram of a fire that spread from a house to grass and then to some trash that was stacked outdoors. Note the broken window, which allowed embers to fall onto the grass, and the separate burn patterns on the grass and on the exterior wall of the house. (The investigator would also have photographed the areas shown here and in Figure 6-8, in support of his sketches.)

Trees and Shrubs. Trees or shrubs that are close enough to be affected by the fire can be used to determine its path. Leaves that are green will be turned brown or shriveled by the heat. Branches without foliage (and

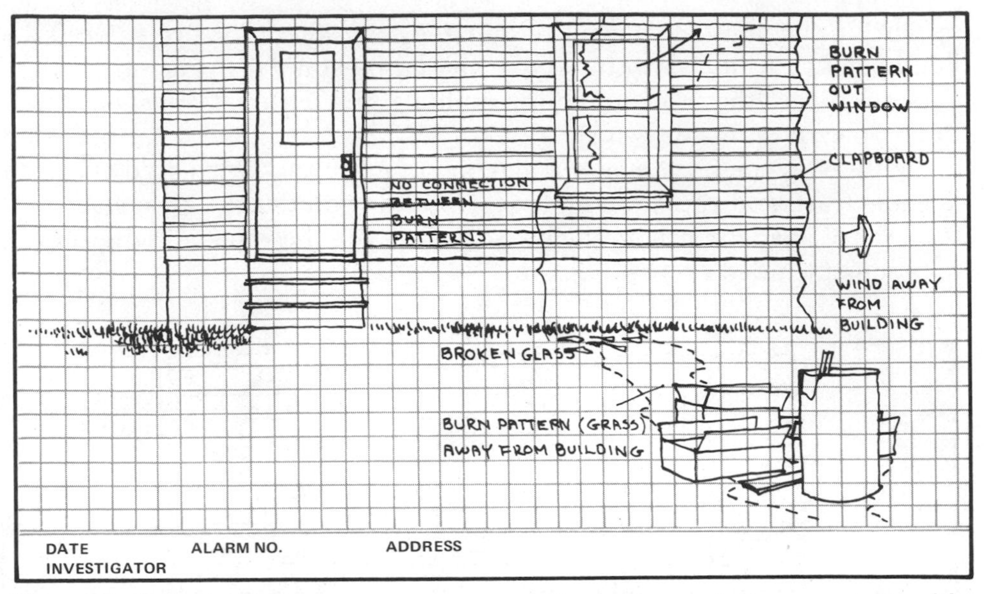

Figure 6-9. Notice that what appears as two unconnected burn patterns is due to communication of the interior fire igniting the exterior fire.

trunks) may be scorched on the side facing the fire, but unaffected on the side furthest from the fire (Figure 6-10).

CASE HISTORY: Tire Tracks

Late one evening in May, fire was reported in a building containing several doctors' offices. Firefighters found one doctor's suite heavily involved; closed and locked doors (including the door to the involved suite) had kept the fire from spreading through the building.

The flame and smoke colors indicated the burning of only the usual combustible materials. All windows and doors were locked. Only two unusual circumstances were observed by firefighters:

Observation	*Evaluation*
• Window in involved suite found shattered on arrival; glass found on floor inside	• Possible illegal entry—or shattered by heat
• Several rocks of tennis-ball size found in involved suite on the floor	• Rocks could have been thrown through windows

The fire was extinguished, and an investigation begun. The examination of the grounds around the building yielded two sets of bicycle tracks, one deeper than the other. The tracks were diagrammed and photographed. Inside, the investigator examined the shattered window glass. The side of each piece of glass facing the floor showed little or no smoke stain; the

Figure 6-10. *Burned leaves indicate fire and wind direction.*

exposed side was coated with stain. This indicated that the window had been broken before the fire started.

By tracing smoke and burn patterns, the investigator established the point of origin as a wastepaper basket next to a desk. The basket was at least 4 feet (12 decimeters) from any available source of ignition. The investigator diagrammed and photographed the area of the wastepaper baskets, and declared the building a crime scene. He ruled as follows:

Point of origin: Wastepaper basket in physician's office
Heat source: Unknown (probably match or lighter)
Reason: Deliberately set
Category: Arson.

The state fire marshal conducted the arson investigation. One witness mentioned that she had seen a youngster filling his bicycle basket with rocks from a nearby railroad bed. She had not seen what the boy did with the rocks. The fire marshal located the boy and his bicycle. The tire treads matched the tread marks at the fire scene, and dirt embedded in the tire treads was of the same composition as that at the fireground.

The boy admitted that he had bicycled to the office building from the railroad bed and pegged a few rocks through the locked window for "something to do." He had then climbed in through the window and looked around for something to steal. Finding little, he had started the fire with a match, climbed out, and bicycled away.

EXAMINING THE EXTERIOR OF THE STRUCTURE

Each exterior wall of the fire building should be examined for signs of fire and of firefighting activities. If the fire building is a detached structure, all four walls should be observed; if the building is attached (a row building), only two or three exterior walls will be accessible. Some attached buildings are taller than adjacent buildings; in such a case, the wall of the fire building above the adjacent roof line must also be examined.

The objective of this examination is simply to discover all existing evidence of the fire's origin and movement, and of the factors that may have affected it. A small fire that is quickly extinguished may leave no signs at all on exterior walls; then there is no evidence to be found. An extensive fire, however, will most likely leave signs, such as smoke and burn patterns, on several walls and at all levels of the building. These signs of fire travel (and signs of firefighting activities as well) are as important to the investigator's determination of the point of origin, heat source, and so forth, as are the signs found within the structure. In some situations, the interior patterns may not make sense until they are combined with the results of the exterior examination.

A Suggested Procedure

Walk around the building once to develop an overall impression of the situation. Make a diagram of each wall, showing the positions (but not necessarily the conditions) of doors and windows. Then make a second tour of the building, examining each wall carefully. Observe, and mark on each wall diagram, any signs of fire travel or firefighting activities, including:

- Smoke stain patterns
- Charring and scorching patterns
- Seemingly unrelated locations showing evidence of fire
- The lowest point at which there is an indication of fire
- Points at which fire extended from the inside to the outside of the structure
- Points of entry of fire into the building
- Whether each door and window is open or closed
- The condition of all window and door glass: broken, melted, smoke-stained, etc.
- Signs of forced entry
- Positions and conditions of gas and electric meters
- Positions and conditions of fire-protection-system controls and connections
- Positions of hoselines and ladders.

In addition, take at least one photograph of each exterior wall, showing the entire wall. Also take close-up photographs of any unusual conditions. Record, in a notebook or on tape, each condition found, and note which diagrams and photos show that condition. (This is particularly important with regard to photographs; close-up photos of fire damage tend to look alike, and may be confused if they are not carefully catalogued.)

Smoke Stain Patterns

Smoke stains are usually heaviest at doors and windows, especially on and above the level at which the fire originated. When the fire is confined to a small area, smoke stains may be the only external evidence that a fire occurred within the building. Then, the stains will generally indicate the location of the fire.

On the other hand, a deep-seated fire that has been burning for some time will produce enough smoke to stain the exterior heavily (Figures 6-11 and 6-12). The smoke will travel wherever it can within the building, and push out through every crevice and crack, as well as around doors and windows. The smoke may also build up within the structure to the extent that it is forced downward, so that it leaves the building below the level of the fire. The widely distributed smoke stains then give little indication of the location of the fire. However, the heavy dark brown stains, especially on windows, are indicative of a deep-seated, smoldering fire (Figure 6-13). This very heavy staining should be accompanied by evidence of extensive ventilation by firefighters.

Figure 6-11. *External staining may indicate inside areas of burning.*

Figure 6-12. *Heavy, external smoke stains may indicate the heaviest burning inside structure.*

Figure 6-13. *Heavy, dark staining indicates a slow-burning fire.*

Charring and Scorching Patterns

A free-burning fire usually leaves visible charring and scorching patterns on exterior walls. These patterns begin at the point or points at which the fire traveled from the inside to the outside of the building. In most cases, these points are at the tops of open doors and windows. Combustible exterior finishes such as clapboards and shingles may become involved and may exhibit charring and smoke patterns; brick, stone, and other noncombustible exterior surfaces will show scorching and smoke stains.

The patterns on exterior walls are influenced by winds and by the velocity of the fire itself. Burn and smoke patterns that are produced in calm air will rise straight up from the point of exit (Figure 6-14). Patterns produced in a wind will lean in the direction of wind travel (Figure 6-15). High-velocity flames can, however, produce charring patterns that indicate flame movement against the wind (Figure 6-16).

Flame Travel Out Windows. Fire may travel out a window if the window was left open or if the window glass was broken before the fire, by heat from the fire or by fire crews for venting. Note whether the window glass is broken inward or outward and whether it is cracked or melted as well.

The burn patterns on the window frame, sash, and walls around the window should indicate whether the flames moved out through the upper or the lower portion of the opening (Figures 6-17 and 6-18). Flame travel out

Figure 6-14. *This pattern indicates that wind was not a factor.*

Figure 6-15. *The wind traveled from left to right creating this pattern.*

Figure 6-16. *This pattern was caused by the flame moving into a high velocity wind.*

the upper portion of the window opening indicates that, inside, the fire spread along the ceiling to the window; the fire then moved out the window at the highest opening it encountered. Fire travel out the lower portion of the window opening indicates that, inside, the fire traveled up from below the window. The fire may have moved up from a lower story; or, it may have moved along the floor to the wall below the window, and then up that wall to the window.

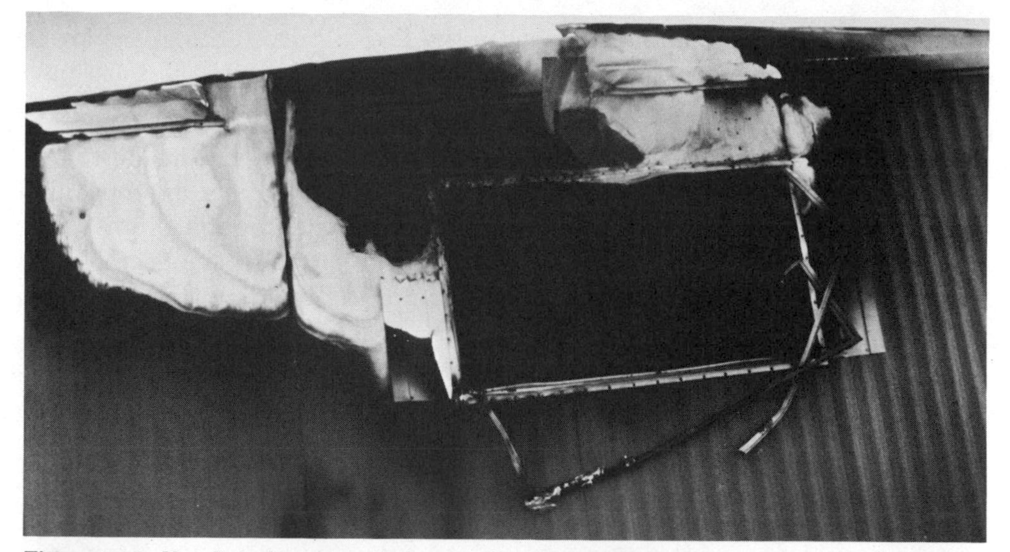

Figure 6-17. *Very little damage to the bottom sill indicates fire traveled across and out the window.*

Figure 6-18. *The damage shown indicates that the fire came from below the window.*

Flame Travel Out Doorways. Flames will usually travel out the top of an exterior doorway, and then only if the door is open. There are, however, two notable exceptions. First, if the fire has been attacking a closed exterior door for a substantial period of time it may burn an opening through the door or break a glass insert in the door; then the path of fire travel through the closed door should be obvious. Second, a pool or trail of flammable or combustible liquid may draw fire through an open doorway at floor level; or the accelerant may cause the fire to burn through the lower portion of a closed door and extend to the exterior. Again, the burn patterns should indicate clearly what took place.

Flame Penetration Through Walls. Fire may move from the inside to the outside surface of a building through such wall openings as vent-duct outlets, fan outlets, air intake grills, and poorly sealed openings for pipes and electric wires. The fire can also penetrate through small cracks in walls due to poor construction or to settling of the building (Figure 6-19). If the fire has actually penetrated through a wall, the exterior smoke and burn patterns will begin somewhat above ground level; there will be an area that is free of charring between the ground and the lowest point of the pattern. (Compare the burn pattern in Figure 6-19 with that on the exterior wall in Figure 6-8, which shows extension of a ground fire to an exterior wall.)

Figure 6-19. *Fire extends through cracks to exterior walls.*

Exterior Burn Patterns as Indicators

Exterior burn patterns are usually excellent indicators of the fire's original location. The doors and windows with the heaviest charring and scorching patterns are most often those that are closest to the point of origin. For example, a fire that starts in a kitchen will probably first travel out the kitchen windows or door (if the kitchen has an outside door), and these will show the heaviest exterior damage. Of course this is not always the case, especially if the fire originates in a location with minimum fuel and then spreads to an area containing much flammable material. However, it is true often enough to be considered seriously by investigators.

Some exterior burn patterns may be the only obvious indications of interior fire spread. An example might be a burn pattern that begins at the lower portion of an open upper-story window. The fire may have burned upward from the story below, inside the wall, leaving no easily visible signs of travel until it vented itself at the window.

Lowest Burn Pattern. The lowest exterior burn pattern should be no lower than the level at which the fire originated. If it is below the point of origin, the investigator must determine why the fire extended downward.

Multiple Patterns. Normal fire extension through a building can produce several unconnected exterior burn patterns. These may be evident on adjoining or opposite walls and at the level at which the fire originated or at higher levels. However, interior burn patterns should show the connections among the several exterior patterns—that is, the path of the fire from the point of origin to the location of each exterior burn pattern. The lack of a reasonable path of fire travel may indicate that there was more than one point of origin.

Stacked Patterns. Exterior burn patterns may be stacked one above the other at several building floor levels. This usually means that the fire traveled upward through a stairway or other vertical shaft, entered a room at each floor, and extended out a window or other opening. The stack of burn patterns may skip a floor or two; for example, burn patterns may show on the second, third, and fifth floors, but not on the fourth. This simply means that the fire was unable to extend out of the shaft at the skipped (fourth) floor.

Leapfrogging. Leapfrogging is the mode of travel in which fire extends out one window, up, and back into the building through the window directly above. It can occur at any vertical row of windows, and can account for seemingly separate fires at the same location on several floors of the fire structure (Figure 6-20). The exterior burn patterns produced by leapfrogging can be easily distinguished from the stacked patterns produced by upward fire travel via vertical shafts.

In another form of leapfrogging, fire travels out a window but re-enters the building through a vent, such as those placed in soffits to ventilate attics or cocklofts (Figure 6-21).

Utilities and Fire-Protection Systems

Utility connections and fire-protection-system connections and controls may be damaged to hinder firefighting or fire investigation efforts:

- A gas-meter cock may be intentionally damaged so that it cannot be closed, to allow gas to feed the fire (Figure 6-22).

Figure 6-20. *Exterior burn patterns produced by leapfrogging.*

- The supply of electricity may be shut off at the meter to disguise the time at which the fire actually started (by stopping all electric clocks earlier than they would have been stopped by the fire).
- On a standpipe system, the fire department siamese connection or the valves at the hose connections may be damaged to prevent the pumping of water into the fire building; on a sprinkler system, the fire department siamese may be damaged for the same reason.

Obviously, any of these conditions are cause for a careful search for evidence of arson inside the fire building—and perhaps reason enough to categorize the fire as arson.

EXAMINING THE ROOF

The roof should be examined for the purpose of determining whether it was involved and, if so, how. It may be examined before or after the interior of the building is investigated. Most investigators prefer to leave the roof for last, unless the interior fire was confined to the attic or cockloft.

Fire can be carried directly to the roof of a building by convected embers or by vertical shafts that terminate on the roof. Fire can also burn up through the roof from the floor below or from the attic. An examination of the affected roof area, both from above and from below, should allow the investigator to determine which was the case.

Obviously, if the roof was involved but the fire did not burn through the roof, then it must have originated on the roof. However, in some situations

Figure 6-21. *Fire reenters building through roof cornice vents back into the attic.*

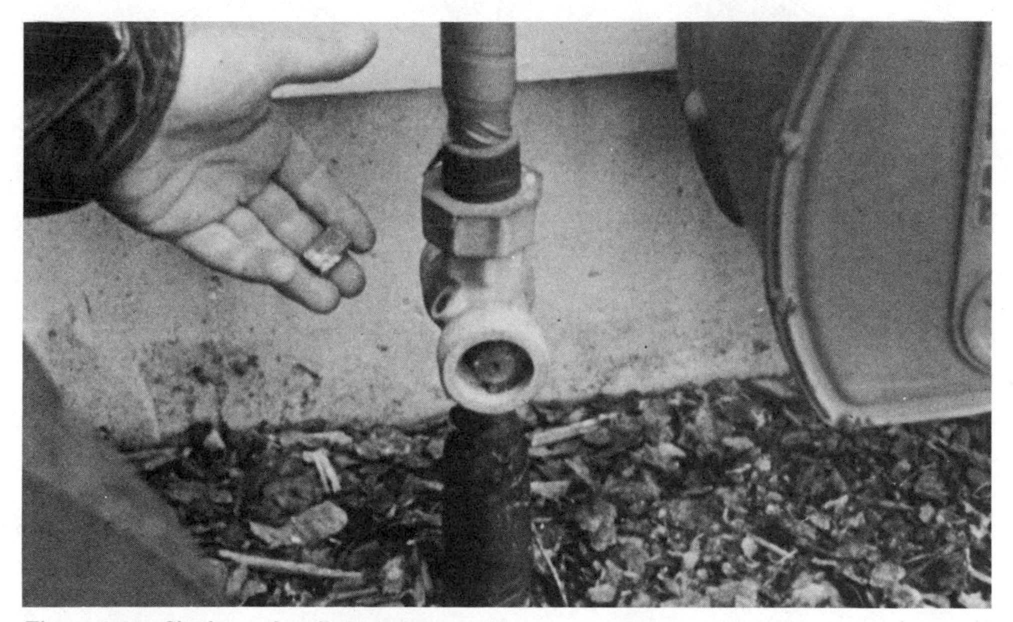

Figure 6-22. *Check gas shutoffs for intentional damage.*

fire may involve both the roof and the attic area, leaving a hole in the roof. The investigator must then determine whether fire in the attic burned up through the roof, or fire on the roof burned down into the attic. (The absence of burn patterns on the floor below the attic may mean little here, since fire could reach the attic through a vertical shaft or by leapfrogging.)

A fire that originates in (or travels into) an attic will burn upward until it is stopped by the roof. It will then mushroom, creating severe burn marks along the interior surface of the roof. Eventually, the attic fire will burn through the roof, venting itself. The fire will then be pulled up through the roof opening, involving the roof (Figure 6-23a). The smoke and burn patterns will show the mushrooming and the movement of the fire toward the roof opening. The edges of the opening will also indicate the direction of flame travel (edge burn patterns are discussed in Chapter 7).

A fire that originates on a roof will take a very long time to burn down through several layers of roof material. During that time, the fire will do a great deal of damage to the roof itself—much more than fire penetrating upward from the floor below, and over a much larger area. Once the fire burns through and falls to the attic floor, it will burn upward and vent itself through the roof opening; this will limit the fire spread and fire damage in the attic (Figure 6-23b). In addition, the edges of the opening will indicate the direction of flame travel.

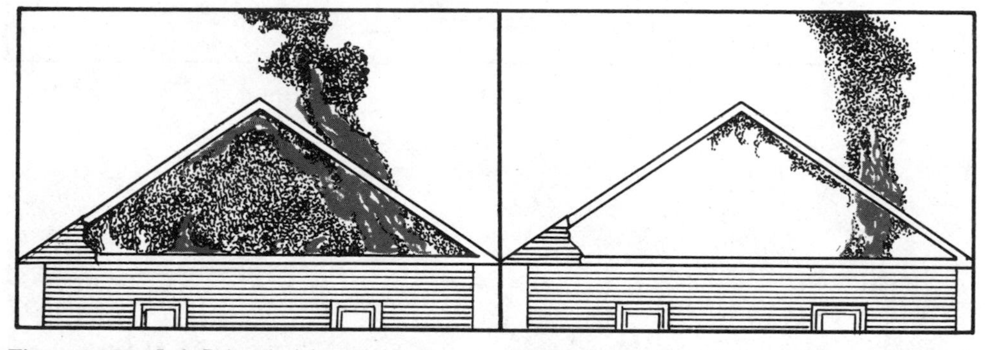

Figure 6-23a. *Left: Point of origin: attic interior.* ***Figure 6-23b.*** *Right: Point of origin: external roof.*

CASE HISTORY: Roof Fire

On a warm Saturday afternoon in June, two engine companies were dispatched to a reported garage fire. Upon arrival, the first due company noted fire in the cockloft as well as fire on the outside of the roof. While attacking and extinguishing the fire within the attic, the hose crew noticed that the fire was burning from one area of the cockloft floor upward to a hole in the roof. The crew attacking the fire on the exterior of the roof stated that the burned area was large, and it appeared that the fire had burned down through the roof.

Upon checking the surrounding area, the fireground officer noticed a barrel that was used for burning trash. The barrel had a fire in it and was downwind of the garage. The air holes in the cover of the barrel were very large. The officer concluded that embers from the fire in the barrel had been carried by the wind to the garage roof, which was 20 feet away. After the roofing material was ignited, the fire had burned a hole in the roof and extended to the inside of the garage. His report read:

Point of origin: Exterior roofing material
Heat source: Embers from barrel fire
Reason: Improperly covered barrel, too close to garage
Category: Accidental.

7

INVESTIGATING THE FIRE: BUILDING INTERIOR

The object of the interior examination is to determine—or to find clues that will help determine—the point of origin, heat source, reason, and category of the fire. If the fire is of natural or accidental origin, the clues will consist of normal heat, burn, and smoke patterns that lead to the low burn. If the fire is an arson fire, there may be, in addition, erratic burn patterns, the remains of arson devices, such unusual conditions as signs of forced entry or abnormal contents, or more than one low burn. The investigator performing the interior examination must use his knowledge and observation to find these clues, to realize that they are clues, and to decide what they mean. He must then correlate information gathered during the interior examination with information gathered during the other parts of the investigation, to make his determinations concerning the fire.

In the first section of this chapter we provide a procedure for conducting the interior examination. The procedure is deceptively simple and can be summarized in a single sentence: *Start at the point of least damage and work your way back to the point of most damage, observing everything along the way.* The implementation of the procedure, however, can be far from simple. For that reason, the remainder of the chapter deals with conditions that may be found during an interior examination and the meanings of those conditions.

SAFETY

Fire damage and firefighting activities may weaken the fire st overload it, and create hazards such as hanging wires, pipes, and s components. Floors may be littered with broken glass and wood debris with protruding nails. Holes may have been burned through floors or cut through for venting. Before entering the building, the investigator should check with the officer in charge, to ensure that all parts of the building are safe.

If the fire is in progress when the building is entered, the investigator must wear full protective gear. If the investigation is conducted after extinguishment, he should wear at least a helmet and boots. In addition, for safety and for good visibility during the examination, he should carry a powerful handlight.

A SUGGESTED PROCEDURE

From the time he takes his first step into the fire building, the investigator must concentrate on the job at hand. And the essence of the job is observation. For example, as he moves in through an exterior doorway, the investigator should be examining it for signs of forced entry. As he moves toward the involved areas of the building, he should be checking the condition of windows and interior doors; looking for signs of smoke or heat damage; listening to the comments of firefighters on the scene (eavesdropping might be a better word); noting the placement, condition, and quantity of contents; and seeking anything that might be out of the ordinary—even in seemingly uninvolved parts of the structure.

Here, and throughout the interior examination, the investigator should make notes, either in writing or on a tape recorder, of everything he finds. Anything unusual should be diagrammed and photographed as it is found. Evidence should be marked as such to keep it from being moved or damaged before it is collected; the investigator should have equipment for this purpose (see Chapters 6 and 8). Information gathered from conversations with firefighters who are in the building should be recorded.

At a fire of any size, the investigator will be surrounded—and perhaps bombarded—by information during his inspection of the interior of the fire structure. He must take the time to sort out this information, record it, evaluate it briefly, and allow it to suggest further avenues for investigation. If he tries to absorb it all at once, he may quickly become lost in a maze of burnt debris. The best way for the investigator to control this flow of information is to establish a routine—to proceed in the same way during every interior examination. The procedure outlined here, or a routine developed by the investigator through experience, will help ensure that nothing is overlooked.

Assess the Situation

Before starting the detailed examination, look over the area that was involved. Note, approximately, the area or areas of heaviest fire damage, the seat of the fire, and the direction or directions in which the fire seems to have spread. Try to get a "feel" for what might have occurred. Look for possible paths of fire travel, including contents, finishing materials, horizontal or vertical shafts, and perhaps trailers. Do not, however, come to any conclusions concerning the origin or spread of the fire; simply assess the situation as it is and as it might have been before and during the fire.

If the fire was confined to one room or one area, this initial assessment (and, in fact, the remainder of the interior investigation) should be accomplished with a minimum of difficulty. However, it will not be so easy if the fire was quite extensive and, for example, the remains of the second story are scattered about the basement and first floor of the building. Nevertheless, such extensive fires can (and have been) investigated successfully—even when the investigator must mentally reconstruct the building to determine what took place.

Find the Low Burn (or Burns)

Locate the area of least fire damage; its location will depend on how the fire spread from the point of origin, but air currents and drafts usually push fire and smoke (and their damage) more in one direction than in others. Smoke stains and heat and flame patterns should be visible on walls and ceilings. The patterns should become heavier and/or darker and lower as they approach the low burn (Figure 7-1). Follow the downward slope of the patterns and the increasing severity of the burns and stains to the area of heaviest damage. The low burn should be in this area. (Burn patterns and their meanings are discussed in much more detail later in this chapter.)

While tracing the burn patterns back to the low burn, note the depth and severity of charring, scorching, and blistering on interior finishes and contents. If the patterns indicate a flame intensity or velocity that is abnormally high, record that fact and photograph the abnormal patterns.

Figure 7-1. *Burn pattern narrows and darkens as it approaches the low burn.*

Look for other indications that an accelerant may have been involved. Look for the remains of arson devices and, generally, for unusual, meager, or misplaced contents.

Locate the low burn (Chapter 3). Diagram and photograph the low burn and the path of the fire. If there are other low burns in the structure, repeat the procedure for each of them. Determine which is the true low burn by associating a heat source with the point of origin at the true low burn.

Determine the Heat Source and Point of Origin

If only one low burn is found, make sure it is not simply the visible evidence of a true low burn that is hidden from view. Then, if it is the true low burn, determine what combustible material produced that low burn when it was ignited. In other words, find the point of origin of the fire.

Next, note any heat sources in the area of the low burn and point of origin. Determine which heat source (if any) is directly related to the point of origin. That is, find the heat source that was capable of transmitting enough heat to the combustible materials at the point of origin to ignite those materials and, thereby, produce the low burn. These three elements—heat source, point of origin, and low burn—must be identified and associated with each other if the investigation is to be concluded. If they are identified, the heat source and point of origin should be diagrammed and photographed.

If no heat source can be associated with the other two elements, the fire may have been started with a match or some other "portable" heat source. Or the heat source (perhaps a television set or an electric motor) may have been removed by firefighters. A missing point of origin or true low burn may, likewise, be in a piece of furniture that was moved or removed during fireground operations. The investigator must, of course, determine whether or not this was the case. Contents that were moved during firefighting operations may have to be put back into position before the low burn, heat source, and point of origin are determined, or to establish the continuity of the burn patterns.

If two or more low burns are found, determine whether some of them may have been caused by convected embers. Determine whether any of these low burns can be associated with a point of origin and a heat source. Finally, determine whether there are actually two or more true low burns, or if there is other evidence of arson.

Determine the Reason for the Fire

Every structure contains many possible heat sources and points of origin for a fire, but the vast majority of them never combine to produce ignition. Thus, there must be a reason why a particular heat source caused the ignition of a particular combustible material at a particular point. This is the reason that is reported by the investigator. For example, bread does not normally ignite in a toaster and burst into flames that extend to kitchen cabinets and then to the kitchen walls and ceiling. However, that has happened when the toast ejection mechanism failed to operate, so that the bread was heated beyond its ignition temperature. Then the heat source was the toaster heating element, the point of origin was the bread in the toaster, and the reason was a faulty toaster ejection mechanism.

Once the heat source and point of origin are found and related to each other, determine the reason for the fire—how or why the two came together to cause ignition. In the case of the faulty toaster (and in many other situations), expert help may be required. Even though the category of the fire seems obvious at this point, the investigation should not be concluded until the specific reason for the fire is known.

Determine the Category of the Fire

Recall that a natural fire is one that is ignited without human intervention; an accidental fire is one that is caused by human carelessness; and an arson fire is one that is set for malicious or fraudulent purposes. The reason and category of the fire may be determined at the fire scene or some time later, depending on the circumstances.

Using all the observations made at the fireground, determine the category of the fire. If the fire is of natural or accidental origin, collect and preserve the evidence as detailed in Chapter 8; conclude the investigation with the required reports (see Chapter 11). Evidence of a natural or accidental origin is preserved for use in the event that a civil suit arises from the incident.

If the investigation does not produce enough evidence to label the fire as natural or accidental, or if any evidence seems to indicate a suspicious origin, proceed as follows:

1. Categorize the fire as arson.
2. Establish the fireground as a crime scene according to standard procedures.
3. Advise firefighters to protect all evidence.
4. Notify the cognizant fire department personnel.

If arson investigators (police or fire marshals) cannot reach the scene within a few hours, the fire department may have to retain custody of the fireground. The fire investigator may also have to perform a detailed search for evidence of arson and collect, preserve, and maintain custody of whatever evidence is discovered. The search for evidence and the procedures used in handling evidence are discussed in Chapter 8.

CASE HISTORY: Mobile Home Fire

Rural fire companies were called to an early morning fire in a mobile home, at a considerable distance from the fire station. The response time was 25 minutes. When firefighters arrived, the home was completely involved with intense fire; flames of various colors were roaring out of the doorways and windows, and around a collapsed section of the roof. Since it was impossible to enter the structure, the fire was extinguished from outside. The collapsed roof section had to be removed with a borrowed farm tractor.

After extinguishment, the fireground officer surveyed the scene. A large part of the roof was gone, and much of the aluminum skin on the outside walls had melted. The inside walls were almost completely consumed. It seemed that all that was left was the floor or platform of the mobile home, covered with debris. The officer felt that the fire was beyond his ability to investigate; he requested and received the help of a trained investigator from a nearby city fire department.

The investigator learned little from the observations of firefighters who were at the scene. However, he was able to find a number of clues to the origin of the fire. Melted plastic and deep char patterns on the remains of wooden furnishings showed the direction of flame travel. By backtracking (in the opposite direction of the fire travel), the investigator came to a hole in the living room floor. He assumed that the hole either was the point of origin or was very close to it. The aluminum siding confirmed his assumption: Near the hole, the siding had melted down to the floor, but further away from the hole, some siding remained; the siding that was still intact reached higher as it got further from the hole. At the furthest corner of the mobile home, the siding was intact from the floor to a part of the roof that had remained in place.

The investigator examined the hole in the floor and found that the charring around its edges was typical of downward fire travel. The remains of a sofa, almost directly above the hole, were also examined. Almost nothing was left of it but some charred wood and the sofa springs. Yet it was obvious that the most intense burning had occurred on the part nearest to the hole in the floor. Also, the springs nearest the hole were "dead," whereas those farthest from the hole still retained their springiness; this indicated that the springs near the hole had been subjected to the most intense heat.

Through questioning, the investigator learned that both occupants were heavy smokers. Further, they said they had been watching television in the living room on the evening before the fire and, as usual, had been smoking. The investigator ruled as follows:

Point of origin: North end of living room sofa
Heat source: Cigarette
Reason: Careless smoking
Category: Accidental.

READING HEAT, FLAME, AND SMOKE PATTERNS

Once a fire starts, the hot combustion products tend to rise and move laterally outward from the point of origin. It is this upward and outward movement that extends the fire and produces the V-shaped pattern so important in fireground investigation. In perfectly still air, the V is fairly symmetrical, extending about equally to both sides of the point of origin. A draft of any sort will push the combustion products to one side, so that the V-shaped pattern leans or tilts in the direction in which the air is moving.

The damage is heaviest at the point of origin because the fire has burned longest there. Since it takes time for the fire to travel, involved materials that are further from the point of origin have burned for a shorter time and exhibit less damage.

These two aspects of fire damage—the shape of the pattern and the decreasing severity—allow the investigator to trace back from the area of highest and least damage to the point of origin. Along the way, the investigator can learn other facts concerning the fire by carefully observing the objects that were involved.

Each object involved in a fire is affected in a more or less unique way by heat and flames. The type of damage done and the rate and intensity of burning depend on such factors as the material, shape, surface area, and exterior finish of the object. These factors, and the fire itself, determine the patterns of damage that remain on the object after the fire is extinguished. The investigator can "read" the patterns on individual objects to reconstruct the course of the fire and to discover whether or not any abnormal circumstances or substances were involved.

Walls and Ceilings

Burn patterns and smoke stains are usually most evident on walls and ceilings. A fire that originates on a wall will leave the usual V-shaped pattern, with the point of the V at the low burn. A fire that originates somewhere else but extends to the wall will leave a similar V-shaped pattern; the point of the V will be at the point where the wall first became involved, which is, in this case, not the point of origin. If the fire was led to the wall by a trailer, a burn pattern indicating a trailer will be evident.

The width of the V depends on the burning speed (which, in turn, depends on the material that burned). A material that burns slowly exhibits a wide V (Figure 7-2). A fast-burning material produces a narrow V as the fire moves quickly up the wall (Figure 7-2). An accelerant that is splashed on a slow-burning wall material will cause the fire to move quickly, producing an abnormally narrow V and providing cause for suspicion.

A fire that starts within a wall is almost always due to faulty wiring. The wiring then will show signs of heavy involvement. The wall should not be damaged below the level of the wiring. However, the fire within the wall will

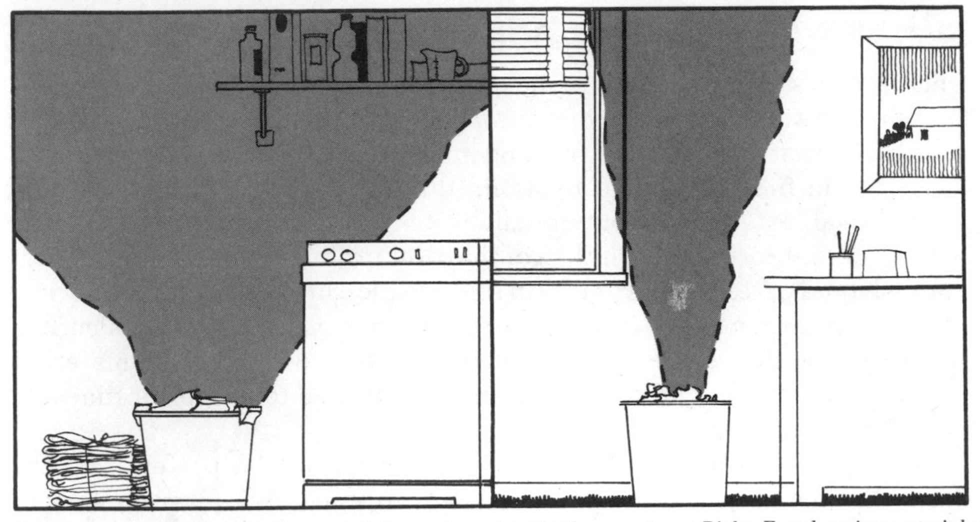

Figure 7-2. *Left: Slow-burning materials create a wide "V" burn pattern. Right: Fast-burning materials create a narrow "V" burn pattern.*

push heat and smoke out around baseboards and moldings. The smoke will leave short, wispy stains that emanate from crevices. The heat will cause paint to bubble and wallpaper to become discolored.

When a fire originates near the center of a room, the flames reach up toward the ceiling where they fan outward. The smoke (and flames, if the ceiling becomes involved) leaves a fan-shaped pattern on the ceiling. If the walls of the room also become involved, the investigator will find both a fan pattern and a V pattern.

Concrete. A room with concrete walls and ceiling will usually contain a fire and its heat, provided the doors and windows are reasonably tight. In some instances, this containment is so efficient that the fire is extinguished because of a lack of oxygen; in other cases, the fire is reduced to a smoldering state, producing explosive, heated gases and a backdraft condition.

It is highly unusual for fire to be transmitted through a concrete wall. If fire is found on both sides of a concrete wall or ceiling (or floor), a continuous burn pattern or path should be evident—through a doorway or a vertical or horizontal channel, perhaps. Otherwise, the fires should be considered suspicious.

A temperature of about 2000° (1093°C) or above will cause concrete to *spall.* In spalling, the intense heat causes moisture in the concrete to expand. The pressure produced by the expanding moisture causes small pieces of concrete to pop off the surface (see Figure 7-9). The presence of spalling thus indicates that the concrete was subjected to a temperature of at least 2000°F (1093°C).

Plaster and Lath. Plaster walls and ceilings will retard the spread of fire. The plaster will not usually burn, but it may collapse when subjected to prolonged high temperatures. The process is similar to the process that causes spalling: Moisture in the plaster expands, creating a pressure within the material. In the case of plaster, however, large chunks are broken off rather than small pieces. Most often, the plaster collapses only above the point of origin or in other areas subjected to intense burning over a long period of time.

A fast-burning, hot fire impinging on a plaster ceiling may cause the plaster above the fire to collapse and fall. This will then provide an opening for fire to extend up through the ceiling (Figure 7-3), and for heat and smoke to vent upward. As a result, the damage to contents will be less severe and less widespread than might otherwise be the case. On the other hand, a slow buildup of heat in a room may cause severe damage to the contents, but have little or no effect on the plaster.

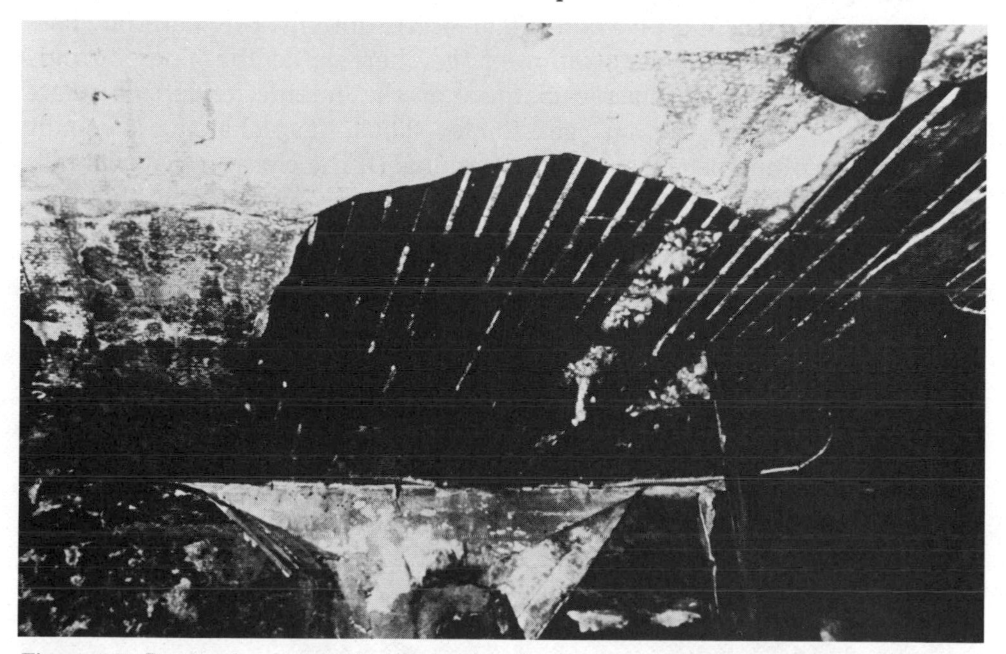

Figure 7-3. *Results of a plaster ceiling impinged by flames.*

Plasterboard. Plasterboard (or dry wall) is a sheet of plaster sandwiched between two layers of heavy paper. It is usually secured to vertical wood studs with nails or screws, to form the interior walls of a room. It may also be secured to wood furring strips or metal studs that have been attached to a concrete wall. In either case, a void space is created behind the plasterboard.

A slow buildup of heat will char the outer layer of paper, but the plaster will usually remain intact. A fast-burning, hot fire will burn the paper away or cause it to curl, exposing the plaster. Then, if the fire is hot enough, the plaster will collapse and fall; this will provide an opening through which the fire can extend into the void space.

Wood Paneling. The wood paneling used as a wall covering is usually a thin sheet of plywood treated with a coating of highly flammable lacquer or plastic. Such paneling burns extremely rapidly, because it has a very high ratio of surface area to volume; it emits flammable gases easily when heated. If heat builds up slowly at or near a sheet of paneling, the sheet will probably bulge out. The coating on the paneling will bubble or burn off. Intense and rapid heating will cause the panel to curl at its edges and then burn quickly, very much as a sheet of thin paper burns.

A more expensive type of wood paneling is composed of either a heavier plywood with a decorative wood skin, or individual boards nailed in place on interior walls. These heavier panels burn like wood rather than paper, producing charring and alligatoring patterns.

Charring occurs mainly in cellulose, or wood fiber, products such as lumber, fiberboard, and plywood (Figure 7-4). Initially, the material chars rapidly, but the rate of penetration of the charring decreases as burning continues. The already charred material on the outside tends to insulate deeper material from the heat and flames. Thus, deep charring is usually an indication of intense or prolonged burning. (If the charred material falls away, it cannot slow the penetration of the charring. However, in that case the remaining char will not be as thick as it would otherwise be.)

Alligatoring is a type of charring that results in a pattern of small lands, or high areas, separated by grooves. The pattern is usually very uniform in appearance. A slow buildup of heat produces a pattern with small square or rectangular lands and shallow grooves (Figure 7-5). Rapid burning and intense heat produce larger oval or round lands (sometimes looking like bubbles) and deep grooves (Figure 7-6).

Figure 7-4. *Charring in cellulose materials.*

Figure 7-5. Results from a slow buildup of heat.

Figure 7-6. Results from rapid burning and intense heat.

An irregular alligatoring pattern (with some small lands and shallow grooves intermixed with large lands and deep grooves) is a strong indication that an accelerant was splashed or thrown onto the wood. Where accelerant soaked into the wood, the alligatoring is typical of fast burning; where there was no accelerant, the pattern shows slower burning (Figure 7-7).

Wallpaper. Wallpaper that is subjected to a slow buildup of heat without direct flame contact chars but remains attached to the wall. The wall appears to have a black coating where the wallpaper was charred. Intense heat causes the wallpaper paste to fail. The paper separates from the wall and is usually consumed by the fire. Small shreds of paper may be found clinging to the wall, provided the wall has not collapsed (Figure 7-8).

Paint. During the early stages of a fire, paint on nearby walls will bubble and blister. More intense heat may cause the paint to liquefy and run down the walls. The liquefied paint can then become ignited and carry fire down the walls. The heavier the coating of paint, the more rapidly it will carry fire downward.

Plastics. Thin sheets of plastic are sometimes used as decorative wall panels or to allow diffused light through walls or ceilings. They will melt or stretch wherever they come in contact with heat or flames. Generally, the path of the heat or flames is precisely defined on the plastic and may easily be traced.

Floors

Floors are almost always constructed of wood or concrete, and often covered with a flammable material. Since fire normally burns upward, floors are usually attacked from below. However, in some situations heat and flames may leave patterns on the upper surface of the floor. This would be the case if an accelerant were poured on the floor, but the presence of

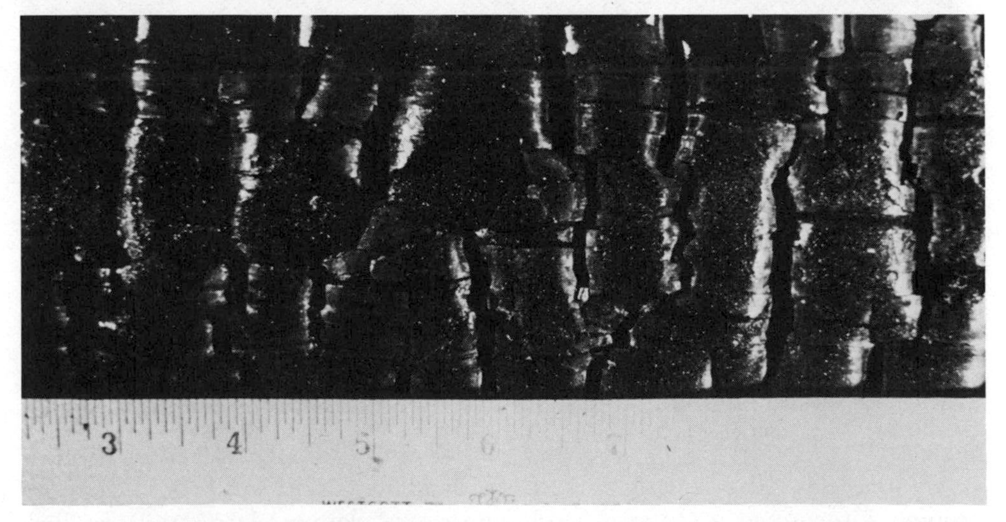

Figure 7-7. *A flammable liquid splashed on wood causes charring of an irregular pattern with deep and shallow grooves.*

Figure 7-8 *Wallpaper subjected to slow buildup of heat chars and remains on the wall.*

a floor pattern does not always mean that an accelerant was used. For example, Figure 7-9 shows the spalling that occurred on a concrete floor when fire originated in a sofa, burned away the carpet under the sofa, and exposed the concrete to heat and flames.

Figure 7-9. *Concrete "spalls" or breaks into small pieces when subjected to heat.*

Direction of Burnthrough in Wood Floors. A hole that has been burned through a combustible floor may be the result of upward burning or downward burning. To determine which is the case, first examine the area below the hole. If the fire burned upward, it would have mushroomed below the floor before burning through. The floor joists below the hole should then show flame patterns that indicate mushrooming.

Next, check the hole itself. If the fire burned from the top down, the upper surface of the floor should be charred over a fairly wide area surrounding the hole. In addition, the top of the hole should be wider than the bottom, so that the edges of the hole form a V in cross section (see Figure 7-10a).

If the fire burned up through the floor, the top of the hole would be narrower than the bottom; the edges of the hole would then form an inverted V in cross section. If the fire continued to burn for some time after burning through from below, the top edge of the hole would be partially consumed. The hole would then appear as in Figure 7-10b: widest at top and bottom and narrowest near the middle.

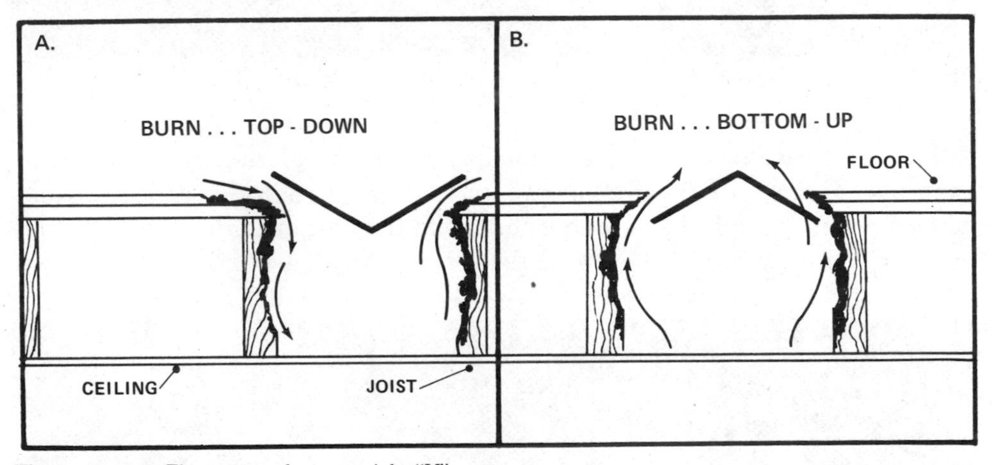

Figure 7-10a. *Fire pattern forms upright "V" pattern.*
Figure 7-10b. *Fire pattern forms inverted "V" pattern.*

Burnthrough Time. Fire will burn downward through softwood flooring (such as pine) at the rate of approximately 1 inch (2.5 centimeters) for each 45 minutes of burning time. The downward burning rate for hardwood flooring (such as oak) is about 1/8 inch (32 millimeters) per 45 minutes burning time. Variations in the hardness of individual boards will affect the burnthrough time; however, these rates can be used as a point of reference in determining whether burning was natural or accelerated.

CASE HISTORY: Fire Extension Through Flooring

Very early on a cold winter morning, fire companies responded to an alarm at a two-story residence. No flames were evident, but dark gray smoke was oozing out around windows and through cracks in the exterior walls. After venting, firefighters extinguished fire in the first-floor living room and in the second-floor hallway. They found the elderly female occupant in bed on the second floor; she was pronounced dead on arrival at the nearest hospital.

A fireground investigator arrived at daybreak and was provided with the following firefighters' observations:

Observation	*Evaluation*
• No flames showing, but much smoke	• Smoldering fire that lacks oxygen
• Smoke dark gray in color	• Long burn time; no unusual substances involved
• Windows heavily stained by smoke	• Slow buildup of heat
• Fire in first-floor living room and second-floor hallway	• Separate fires?
• Open stairway to second floor not involved in fire	• Unusual, especially with second-floor fire—more evidence of separate fires?
• Hole burned through living-room floor	• Check direction of burnthrough
• Occupant found dead	• Probably asphyxiated by gas and smoke from smoldering fire

Heat, smoke, and flame patterns led directly to the hole in the living room floor. Above the hole, the ceiling was heavily damaged but not burned through; the ceiling patterns exhibited the fan shape typical of rising smoke and flames. The remains of a chair were hanging partway down through the hole. The chair was, however, eliminated as the point of origin, because the edges of the floor hole indicated an upward burnthrough. Examination of the flooring and joists from below (in the basement) yielded the true low burn and point of origin; it also indicated the method by which the fire had traveled to the second story without involving the stairway or burning through the living room ceiling.

The fire started where an electrical cable was stapled (through the cable!) between two wood joists supporting the first floor. The fire traveled horizontally between these two joists, to a laundry chute that drew flames and smoke up to the second-floor hall (Figure 7-11). Because the chute effectively vented the basement fire, no other parts of the basement became involved. The fire did, however, burn through the flooring between the two joists, involving the living room. The investigator ruled as follows:

Point of origin: Electrical cable under first-floor flooring
Heat source: Electricity
Reason: Faulty installation of wiring (staple through cable)
Category: Accidental.

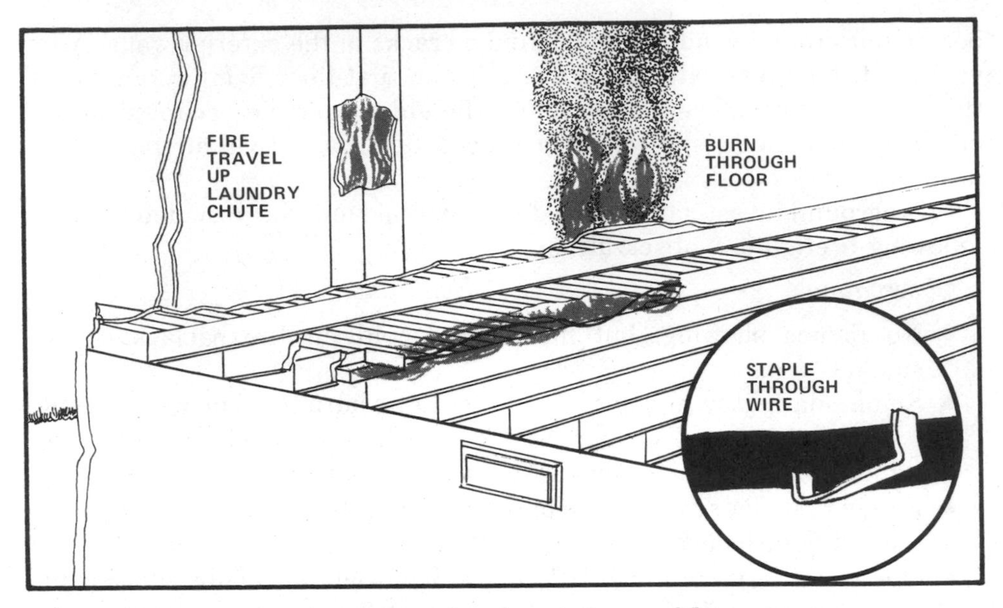

Figure 7-11. Laundry chute drew flames and smoke up to the second floor.

Windows and Window Frames

Smoke Stains. The smoke stains on window glass can indicate the type of fire that produced them. However, the degree of staining will vary; generally, windows that are closer to the seat of the fire will be less heavily stained.

A dense, heavy stain on window glass (Figure 7-12) indicates that the fire built up slowly and did not produce much heat. The burning materials were probably not consumed completely, but they produced a large quantity of smoke containing the products of incomplete combustion. A smoldering fire in an upholstered chair would produce this type of stain. Flammable or combustible liquids involved in a fire may deposit an oily film on glass. This can be detected by rubbing a finger over the surface.

A light stain, or none at all, could indicate that the fire accelerated very rapidly. The fire might produce some stains at first, even quite heavy ones. However, if the stained window glass were then subjected to extreme heat, the stains could be consumed by the fire; the investigator might find only clear glass during his interior examination (Figures 7-13a and 7-13b).

Cracked and Broken Glass. Window glass that has been heavily stained by a slowly accelerating fire may also exhibit the large cracks produced by this type of fire (Figure 7-14a). If the fire eventually reaches a high temperature, large pieces of glass may be broken out of the window; these usually have rounded edges, produced by melting (Figure 7-14b).

A very rapid buildup of heat causes small cracks, fairly close together, in window glass (Figure 7-15). Intense heat can cause irregular, blocklike chunks of glass to fall from the window.

Figure 7-12. *Slow buildup of fire produced little heat resulting in heavy, syrupy stains.*

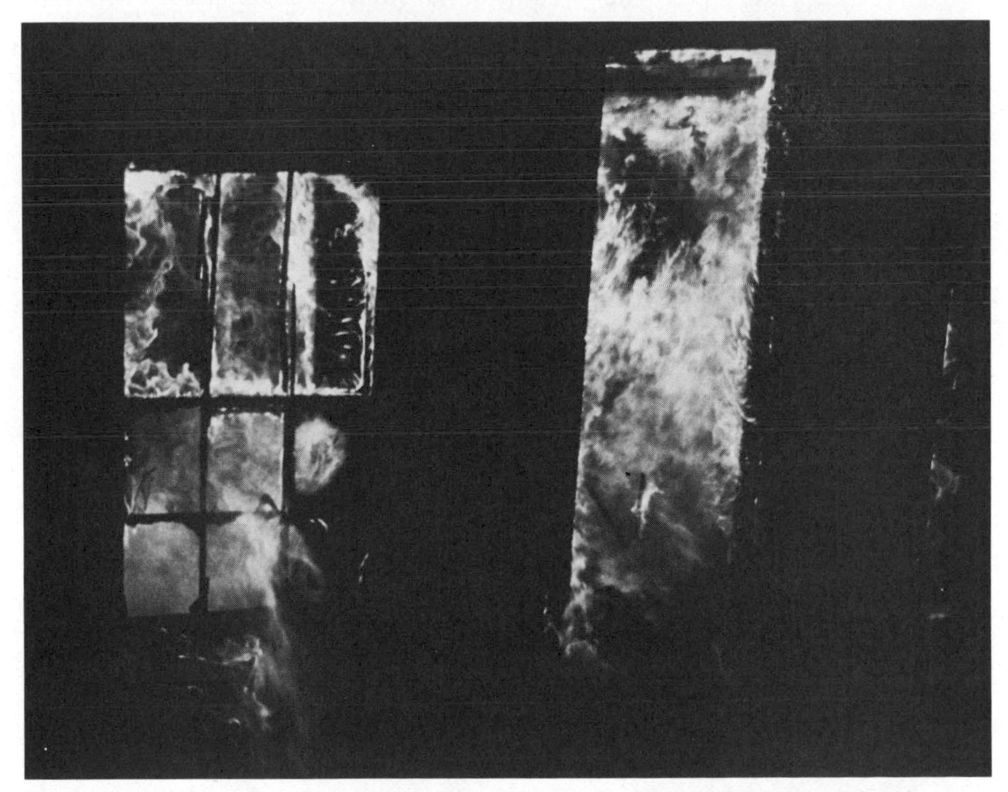

Figure 7-13a. *The stain partially burned off the glass pane but remains on the upper right side.*

Figure 7-13b. *The stain completely burned off the glass pane and melted due to high temperatures.*

When cold water from hose streams or automatic sprinklers or even rainwater hits a hot window pane, the glass will crack. The cracks produced are typically curved in a half-moon configuration (Figure 7-16).

Plastic storm windows will melt, rather than crack, at high temperatures.

Melted Glass. Fires that reach extremely high temperatures—usually fast-burning fires in confined spaces—can melt window glass. In some cases, only the corners of pieces of broken glass may be softened; in other cases, the entire pane may be melted. The melted glass may be completely stained, spotted, or perfectly clear. Melted glass that is found unstained may have been stained during the early stages of the fire. The intense heat of later stages may then have burned off the stains (see Figure 7-14b).

Open Windows. If a double-hung window is open when heat, flames, or smoke reach it, the fire will stain or burn the window tracks. In addition, the window pane nearer the fire will protect the outer pane. If the inner pane is not broken, the difference in staining or cracking will indicate that the window was opened before the fire was extinguished. The investigator must then determine exactly when and why the window was opened.

Interior Doors

A closed interior door presents a barrier to the spread of fire; an open doorway, however, tends to pull the fire through to uninvolved spaces. The investigator should be able to tell, from burn patterns and other clues,

Figure 7-14a. Large cracks caused by a slow buildup of heat.

whether an interior door was open or closed during the fire, and if it prevented or delayed the spread of the fire.

Closed Door. A door that was closed during a fire will exhibit charring and smoke patterns on the side facing the involved space. These patterns will match the patterns on the door frame and surrounding walls. The longer the fire has burned in a confined space, the lower these patterns will extend (due to mushrooming). The edges of the closed door should be free of charring, since they would have been protected by the door frame during the fire. The bottom edge of the door will show charring if a flammable or combustible liquid was poured under it.

Charring on both sides of a closed door may indicate that these were separate fires in the two areas served by the door. An obvious exception is the case in which the fire burned through an interior door of nonresistive construction and extended from one area to the other; then, both the door and the door frame should show heavier damage on the side where the fire originated. Another exception is the case in which there is an avenue for

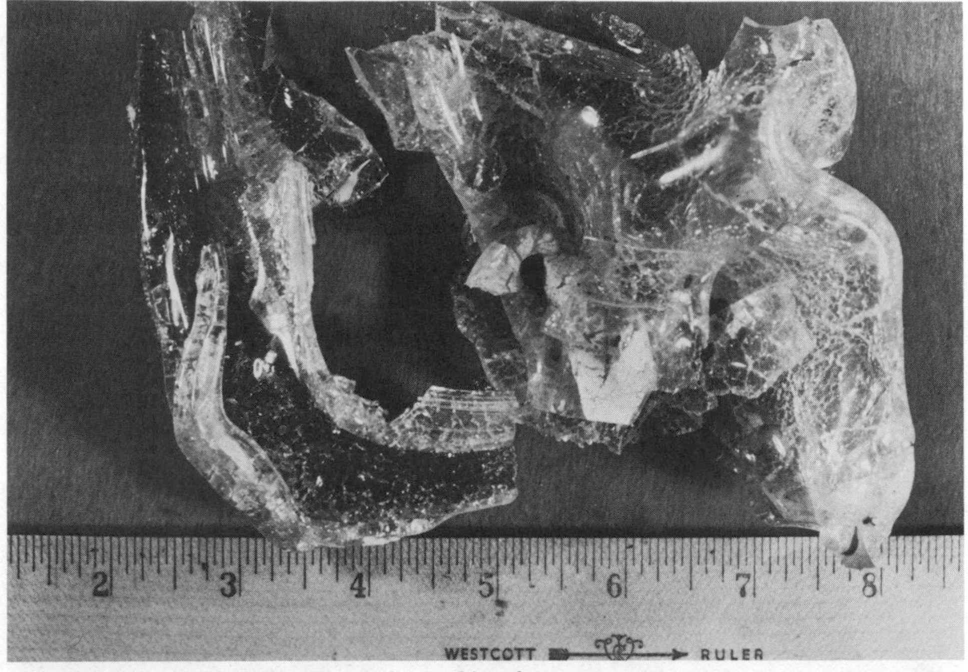

Figure 7-14b. High temperatures result in melting glass.

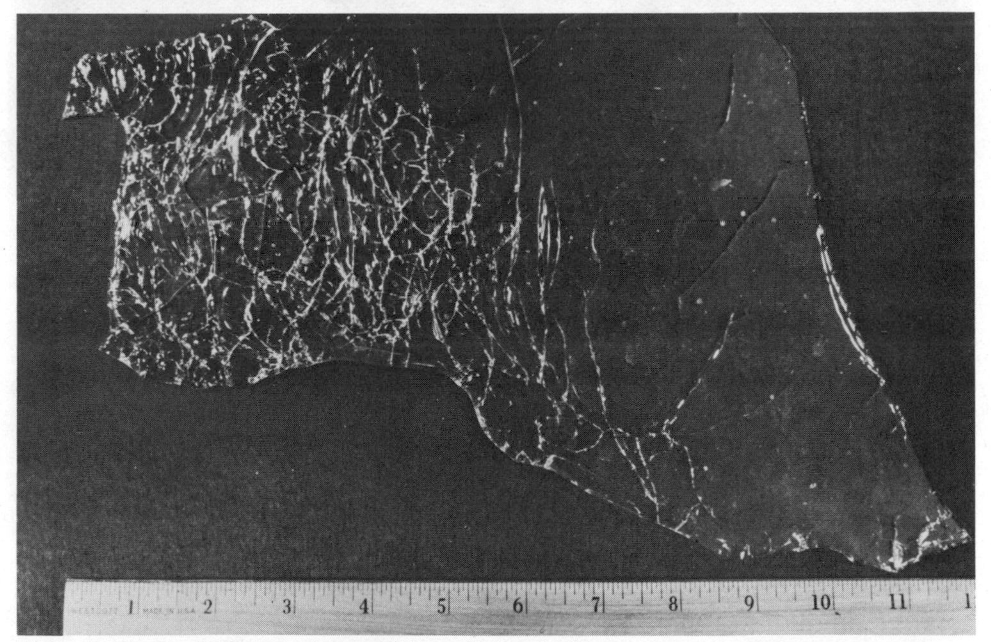

Figure 7-15. Small, close cracks caused by rapid heat buildup.

fire travel between the two areas, such as a floor-level ventilator near the door, or even a wide opening under the door.

Open Door. If the edges of a door are burned, the door was open during the fire (Figure 7-17). The door itself should then be relatively free of smoke

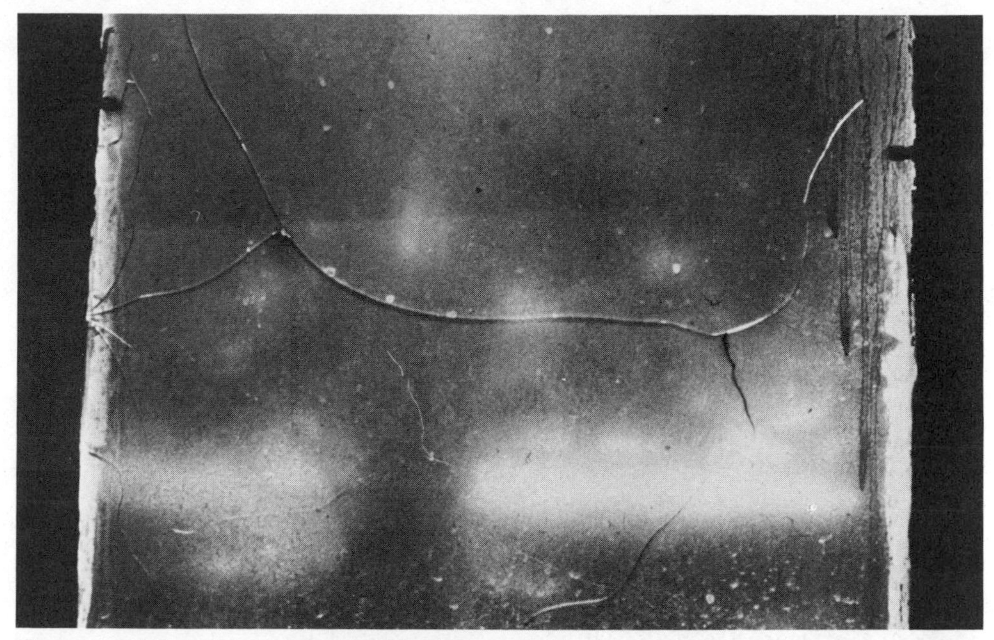

Figure 7-16. *Curved cracks produced by cold water striking hot window.*

Figure 7-17. *Burned door edge indicates door was open during the fire.*

and burn patterns. Instead, charring should be evident on the top (horizontal) part of the door frame. This would result from extension along the ceiling and across the highest part of the open doorway.

Completely Consumed Door. When a door has been completed consumed, the hardware will sometimes provide information as to whether the door was open or closed during the fire. If the doorknob and latch are found next to the door frame, opposite the hinged side, the door was probably closed when it was consumed. If these items are found lying in the room into which the door would normally swing, then the door was probably open. The further they are into that room, the wider the door was open (Figure 7-18).

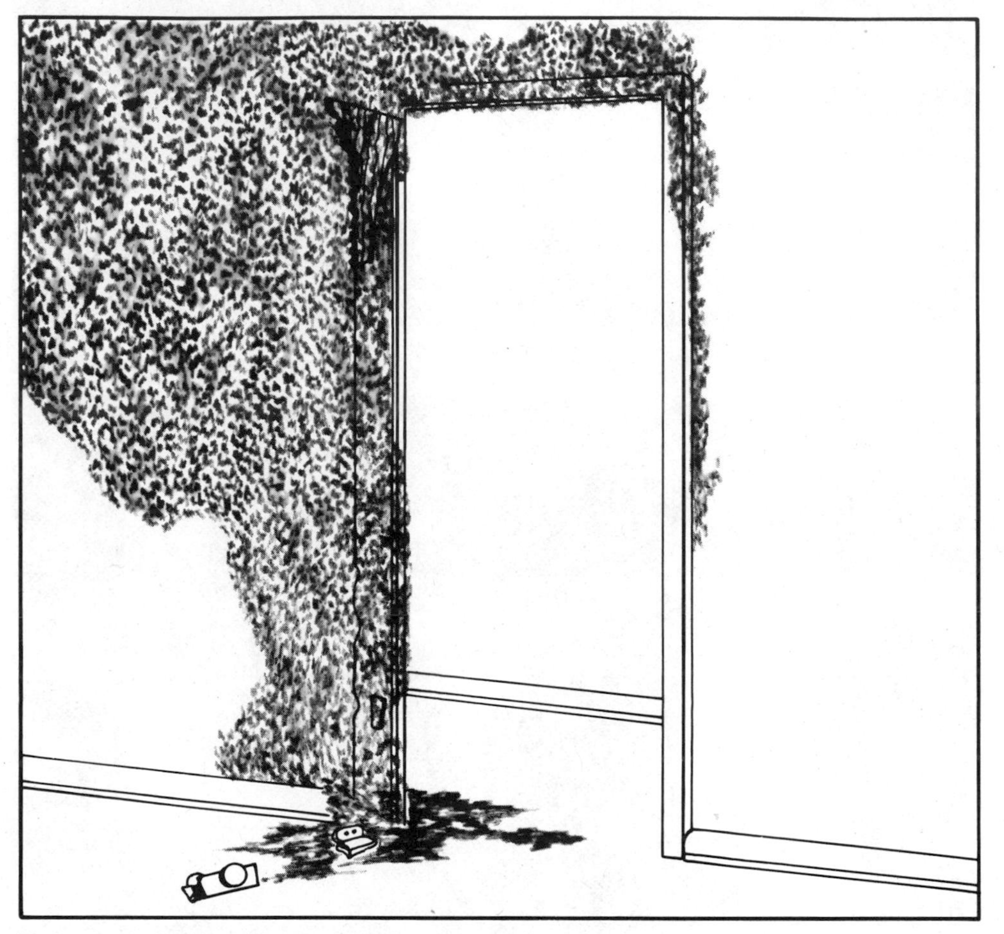

Figure 7-18. *Position of hardware on floor indicates door was open.*

Sometimes the heat will distort the door hinges, so that they become stuck in the position they were in during the fire. This, in turn, will indicate the position of the door at that time. The two leaves of a *mortised* (rectangular) *hinge* lie flat against each other when the door is closed. They open to form an angle as the door is opened; the larger the angle, the wider the door was open (Figure 7-19). The leaves of a *strap hinge* are wide apart when the door is closed; they swing toward each other to form an angle as the door is opened. The smaller the angle, the wider the door was open (Figure 7-20).

Figure 7-19. *Mortised hinge indicates door was open.*

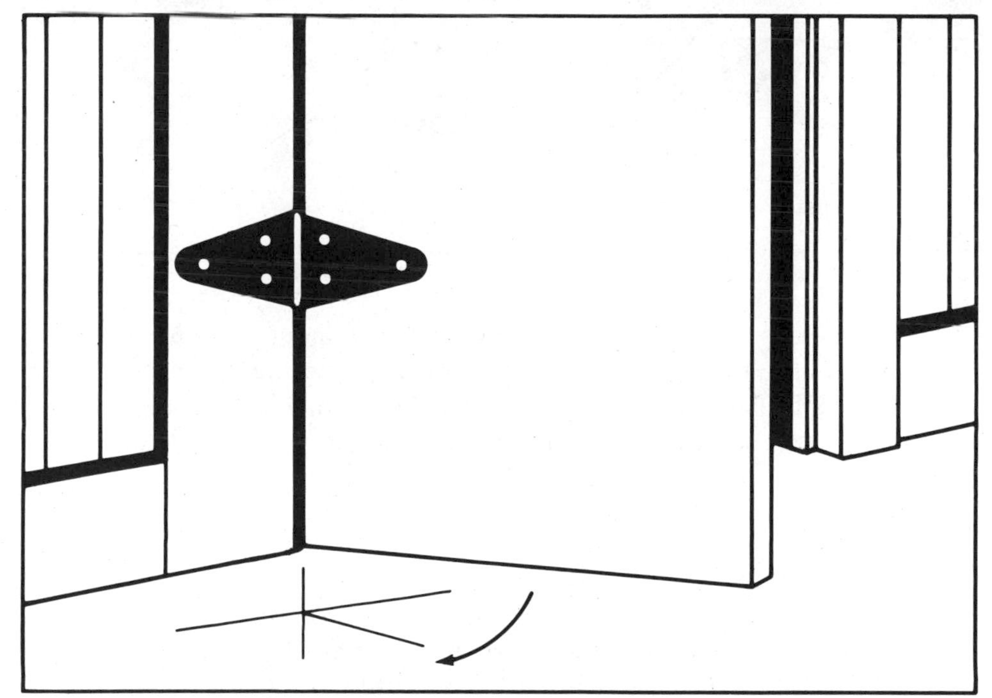

Figure 7-20. *Strap hinges form a smaller angle as the door is opened.*

Varieties of Doors.. The doors discussed so far in this section are the usual wooden doors installed to allow rooms to be closed off, but to permit people to enter or leave. In addition, doors are used to allow access to such equipment as boilers, kitchen cabinets, and dumbwaiter shafts; they are used as scuttles, dampers, and fire curtains in duct systems, and in many other applications. Wooden doors may be solid or hollow or covered with metal. Other doors may be of hollow or solid metal, metal filled with insulation, plastic, or glass. Each reacts to fire in its own way: Metal doors may warp and plastic doors may melt. When a door is a key factor in an investigation, its high-temperature characteristics (including its rated burning time) should be obtained from the manufacturer.

Structural Components

Wood Studs and Joists. Flames impinging on a wooden structural member will burn away more wood on the side facing the fire, and less on the side away from the fire. Thus, as the flames burn through or past the member, they leave distinctive shapes that indicate the direction of flame travel. These shapes are shown in Figure 7-21. In all cases, flame movement is from the wider part of the burn toward the narrower part.

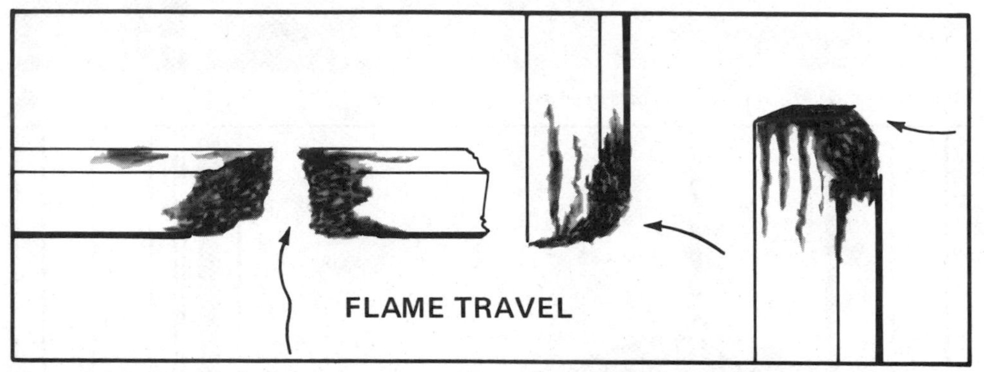

Figure 7-21. *Wood studs burn away more rapidly on the surface facing the fire.*

When charring has been caused by the impingement of flames on wood, the charred area is somewhat shiny. The charring caused by intense heat alone is a dull black. This difference can be used to determine whether or not a charred structural member was actually in the path of the flames.

Metals. At elevated temperatures, metal components may lose their strength or melt. For example,

- At 1000°F (537°C), steel beams will begin to stretch or bend, causing some movement of the walls or floors they support.
- At 1200°F (648°C), steel will lose approximately 60% of its strength.
- At 1500°F (815°C), a structure supported by steel beams can be expected to collapse (Figure 7-22).

At even lower temperatures, steel members may lose their elasticity—that is, their ability to return to their original position after being bent and released. Thus, a hot fire, combined with the structural load, can cause vary-

Figure 7-22. *Steel loses its strength and collapses.*

ing degrees of distortion in steel components. The higher the temperature and the closer a component is to the fire, the greater the distortion will be.

At a temperature of about 1200°F (648°C), aluminum will begin to melt. Aluminum objects closest to the seat of the fire will melt first and most completely, while those further from the fire will show less damage. The melting can therefore be traced back from least damage to most on, say, the siding of a mobile home, as noted in the Case History on page 150.

Building Contents

The contents of a fire building, even when badly damaged, can reveal much information concerning the origin and spread of the fire. Unfortunately, many firefighters view burned or charred contents as useless debris that should be discarded during overhaul and cleanup. Investigators and officers must guard against this tendency; wherever possible, contents and the remains of contents must be left in their original positions until the fireground investigation is completed.

Flames, heat, and smoke leave the same types of patterns on contents as on walls and ceilings. In fact, most burn patterns begin on contents and then extend to walls (that is, most fires originate in the contents). The V-shaped pattern may not be as evident on furniture, for example, as it is on walls—mainly because furniture does not present a large, flat, vertical surface to the fire. Nevertheless, the contents will be marked by contact with heat, flames, and smoke, and the marks will slope downward toward the low burn (Figure 7-23). The V will be wide or narrow, depending on the speed of the fire as it moved across the particular item.

The markings on contents should be consistent with the normal upward and lateral travel of fire. A fire originating in a chair cushion, for example, will normally travel up the backrest and not down to the legs. Fire originating on a workbench might travel laterally along the bench, consuming wood chips and sawdust; but eventually it should move to a source of fuel that will carry it upward.

Surface Patterns. The fire will mark every surface it touches. Lacquers and paints will be discolored, bubbled, or burned. Veneers and laminated surfaces (such as Formica counter tops) will bulge outward where they are subjected to heat. Even the porcelain surfaces of sinks and bathtubs will show traces of scorching and smoke stains. These markings will, however, comprise a smaller part of the overall pattern than the wall and ceiling patterns. For example, only one corner of a chair may be burned, while the rest

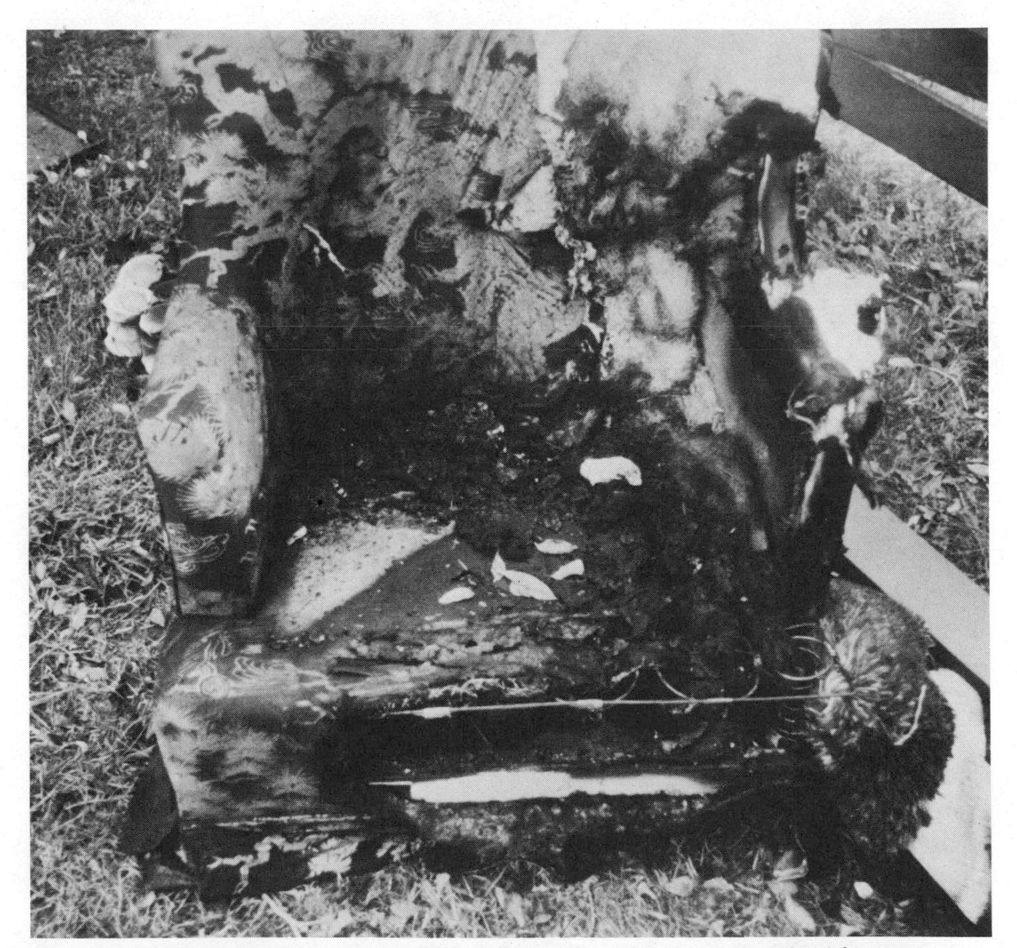

Figure 7-23. *Heavy damage to the right side of chair reveals fire traveled from the right.*

of the chair exhibits only smoke stains; a lamp near the chair may show nothing but smoke stains.

Upholstered Furniture. When flames travel to upholstered furniture, the damage is usually confined to the surface material. The fire will not burn down into the padding, or drop down below the furniture.

When fire originates in a piece of upholstered furniture, it will burn down to and through the padding. As the burning materials within the piece disintegrate, embers will drop down, igniting the flooring or rug below. In addition, the intense heat that builds up within the padding will cause the steel seat springs to lose their elasticity.

To test a spring for loss of elasticity, bend or stretch it by hand (Figure 7-24), and then release it. If the spring remains bent or stretched (Figure 7-25), it has lost its elasticity; the fire most likely originated in the piece of upholstered furniture. If the spring returns to its original shape (Figure 7-26), it is still elastic; the fire must have traveled to the upholstered furniture from some other area. This test is quite reliable; it may be used even if the piece of furniture was completely consumed, so that only the springs remain. It may also be used to determine whether or not fire originated in an innerspring mattress.

Plastic Furnishings. Plastic objects such as telephones and television cabinets begin to melt at about 700°F (371°C). The distortion due to the

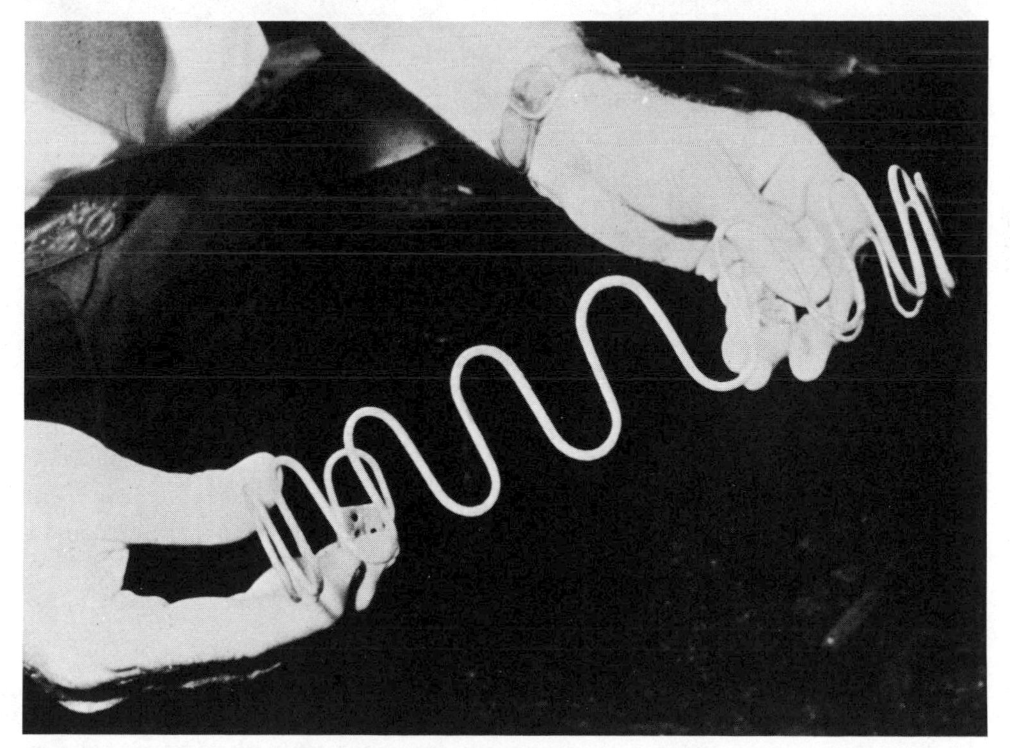

Figure 7-24. *Bend or stretch spring to check elasticity.*

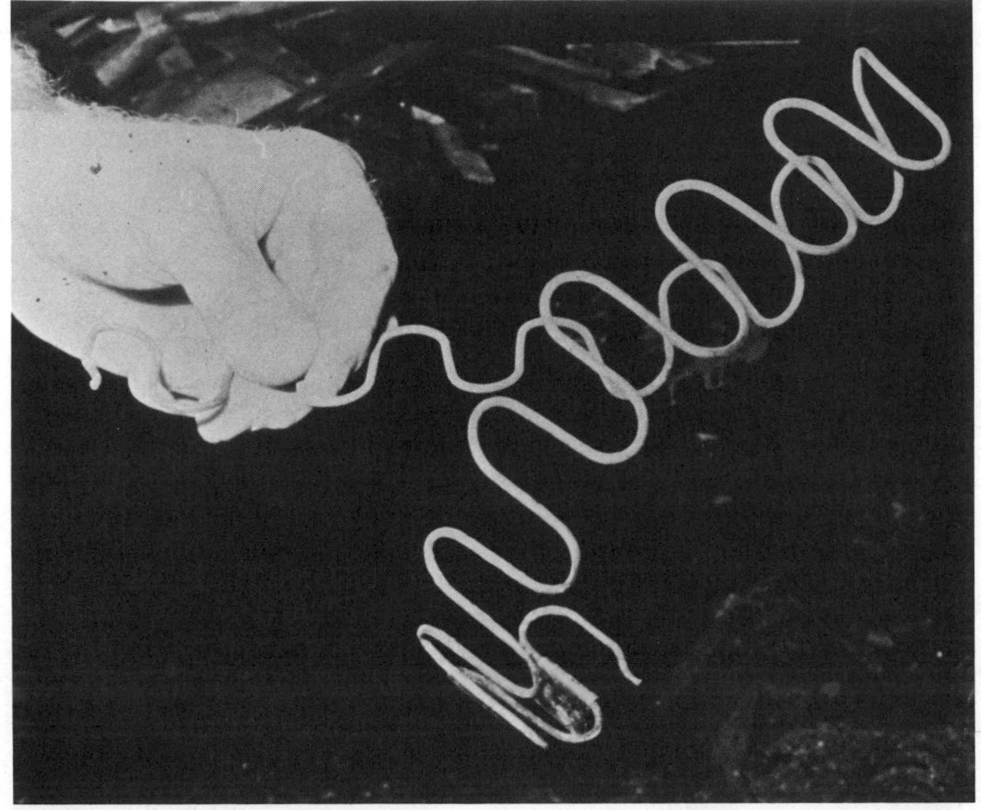

Figure 7-25. *Spring has lost elasticity.*

melting indicates the direction in which the fire was traveling: If the fire was moving horizontally, the side of the object nearer the fire will be most distorted (Figures 7-27 to 7-29). If the fire was moving upward, the bottom of the plastic object will be most distorted; for example, a wall telephone would be melted at the bottom, whereas the top would show little, if any, distortion.

Light Bulbs. When an incandescent light bulb is subjected to a temperature of 900°F (482°C) or more, the glass begins to soften. The melted glass slowly runs down and toward the source of the heat. As a result, a point is formed on the glass; it points in the direction from which the heat approached the bulb (Figure 7-30). This pointer usually doesn't last long enough to be discovered by the investigator, because the cold water from a hose stream will cause the hot bulb to disintegrate on contact. Fluorescent light tubes do not react to heat in this unique manner.

ADDITIONAL ASPECTS OF THE INTERIOR EXAMINATION

The examination of the interior of the fire structure should extend beyond the involved area; and it should encompass more than the tracing of fire

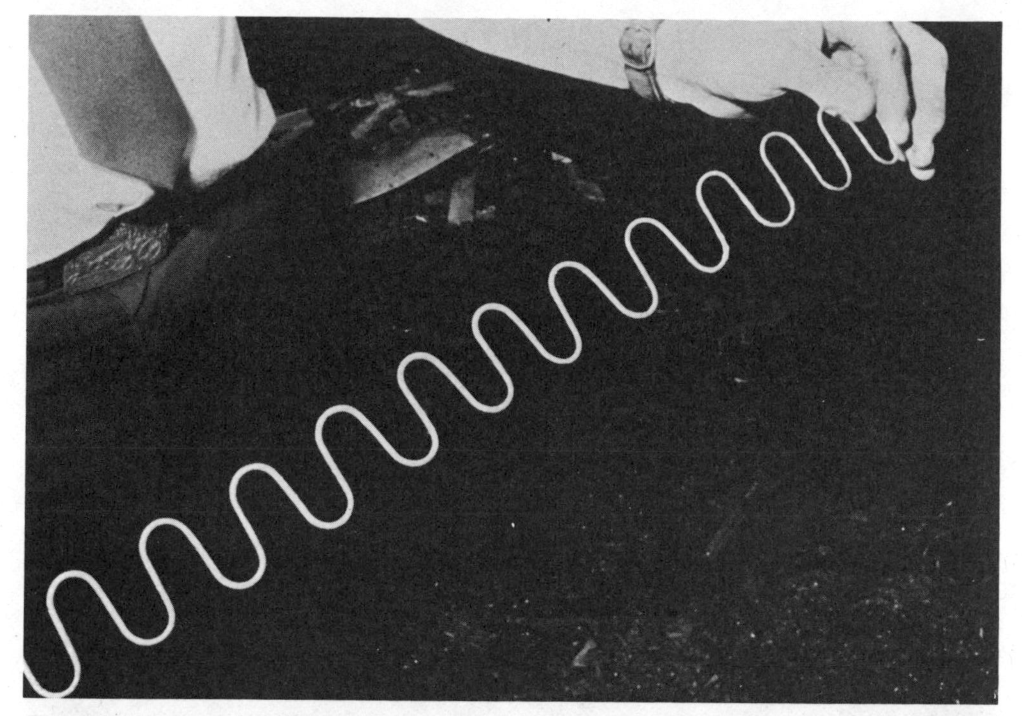

Figure 7-26. *Spring has retained its elasticity.*

Figure 7-27. *Fire moved from right to left melting right side of stereo dust cover.*

patterns. For example, the investigator should make a tour of uninvolved areas—especially those surrounding the involved area—to look for unusual items or conditions. What is or would be considered unusual depends on

Figure 7-28. *Fire moved from right to left and upward distorting the side and bottom of the telephone.*

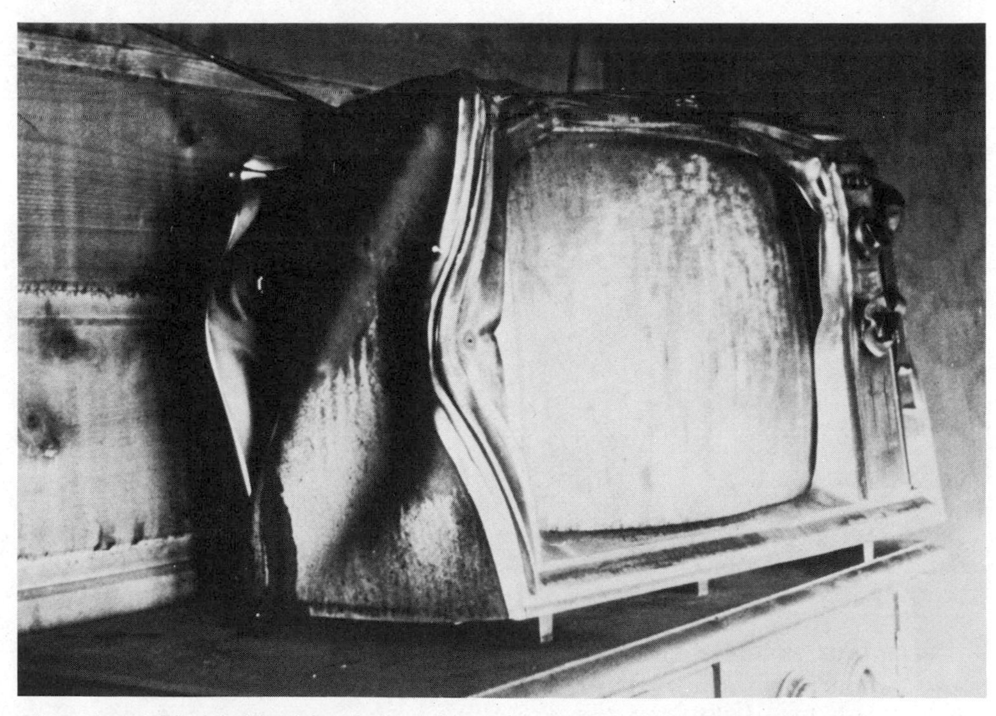

Figure 7-29. *The left side of the television distorted and was drawn toward the fire.*

Figure 7-30. Distorted incandescent light bulb points in the direction from which the fire traveled.

the particular type of occupancy and the occupants. Certainly, an arson device that failed to ignite would be unusual, but most of what the investigator finds will not be so obviously abnormal.

Reconstructing the Scene

Occasionally, the investigator may have to reconstruct part of the fire scene before he can successfully read the burn patterns and smoke stains. This might be necessary, for example, when firefighters have removed burned contents and other debris during cleanup operations. Most of the burn patterns, especially those from the area of the point of origin, would then probably be located in a pile outside a window. Such a situation might at first seem hopeless, but it is not. The important pieces of evidence are available; they must be searched for and found. Occupants and other persons who are familiar with the fire scene may be enlisted to help the investigator reposition contents and even structural components.

In one investigation of a residence fire, the partially burned wall studs that separated the living room and a bedroom were taken back into the structure, repositioned, and nailed in place. The remains of panels from this wall were found and matched to nail holes on the upright studs, and then nailed in place. Once this was done the burn patterns were obvious, and they matched the patterns on contents that originally were adjacent to the wall. In particular, a sofa (also found outside) that was positioned against the wall, was very badly burned. The upholstery was completely gone and the frame was heavily charred, but the back part of the frame showed deeper charring, and the springs had not lost their elasticity. All these signs indicated attack from the space between the sofa and the wall and led to the accurate determination of the point of origin and heat source.

Contents

In both involved and uninvolved areas, the investigator should check for abnormal placement and quantity of contents. We have already noted that there are norms regarding the contents of residential and industrial occupancies; reasonable variations from these norms may not be unusual. However, a condition that is not in keeping with the occupancy or with the purpose of the contents should be scrutinized. Furniture that is stacked in the middle of a living room cannot be used for its intended purpose, but it can be used to start a bonfire. Crates piled in front of an automatic fire door cannot be easily loaded onto a truck for shipment, but they will keep the door from closing in case of fire. A toaster may be found almost anywhere in a kitchen, but it would be very much out of place under a bed. And a retail store with almost no stock would be highly unusual.

Other examples of abnormal placement and quantity of contents have been presented in previous chapters. One area of investigation that has not been mentioned is clothes closets in dwellings. These should always be checked, whether or not they were involved in the fire. Normally, they will be fairly full of clothing, and little else. Even when fire has consumed all the clothing, durable items such as buttons, buckles, zippers, and metal hangers will be found. A completely empty closet should be suspicious, as should unusual contents in the closet (see Figure 7-31).

As a general rule, the investigator should look for contents that are out of place, in disarray, of the wrong type, too meager, or too plentiful for the occupancy involved. Such a condition may or may not be an indication that a crime was committed, and it should not keep the investigator from completing the fireground investigation. However, it should be considered and reported along with other findings—to arson investigators if necessary.

Clocks

Clocks are used throughout most dwellings and places of business, often in every room. The presence of clocks during a fire can indicate the time a fire originated and the path of fire travel. Clocks may be accurate or inaccurate. Each situation will depend on whether the clocks were set accurately and were keeping good time.

A sequence of events during a fire could be established chronologically by several clocks stopping as a result of damage from heat and flames. For

Figure 7-31. Examine closets for durable items which should remain even after a severe fire.

example, fire originating in a kitchen electrical circuit causes the cooking range clock to stop at 11:35 a.m. The fire extends to the living room, causing a wall clock to melt and freezing the hands at 11:45 a.m. (Figure 7-32). Another clock in the bedroom is also damaged and stops at 11:51 a.m. (Figure 7-33). According to the time each clock stopped, it could be assumed the fire traveled from the kitchen (11:35 a.m.) to the living room (11:45 a.m.) and then to the bedroom (11:51 a.m.) over a period of 16 minutes. It could also be assumed the fire originated in the kitchen at about 11:34 a.m. The problem with these assumptions is the difficulty in establishing the fact that the clocks were set at the correct local time, and were in synchronization with each other. People often set clocks 10 to 15 minutes fast, in bedrooms especially. If we add 15 minutes to the bedroom clock in the above example, the time sequence would indicate the fire reached the bedroom before it reached the living room.

Arsonists often change the time on clocks to establish an erroneous time of origin. This gives the arsonist the opportunity to set up his alibi by being seen at another location during the time the fire was supposed to have originated.

Delays in transmitting the fire alarm can also be revealed by comparing the clocks damaged at a fire to the time the dispatcher received the alarm. Delays of 15 minutes or more should be suspect, and the reason for the delay should be ascertained. If the answers given are not reasonable, the

Figure 7-32. *Wax melted on the face of the clock preventing the hands from rotating.*

Figure 7-33. *The face of the clock melted and adhered to the hands.*

investigator should look for other evidence associated with delayed alarms as discussed in Chapter 5.

Clocks can help establish the approximate "time of origin" and the direction of the fire travel, especially when they reinforce the physical evidence of fire damage. However, the time shown on damaged clocks cannot be relied upon as unquestionable evidence in court unless the accuracy of the clocks' local time setting and accurate time-keeping can be substantiated.

Fire Doors

Fire doors that are normally closed, such as those leading to stairways, may be blocked open by an arsonist to enhance fire spread. If any are observed or reported to have been open during the fire, they should be checked carefully. The investigator should look for blocking with nails (see Chapter 5), chocks, building contents, or other items. The self-closing mechanisms should be checked for tampering.

Automatic fire doors, which close when heat melts a fusible link, may be blocked open, made inoperative, or delayed. Any automatic door that did not operate properly should be examined thoroughly. A *gravity* door may be made inoperative by damaging the door tracks (Figure 7-34). A *counterweight* door may be kept from operating by blocking or raising the counterweight, so that it will not pull the door closed.

The operation of an automatic door can be delayed for a long enough time to ensure that fire spreads through the opening. This is accomplished by insulating the fusible link—usually by wrapping it with a combustible material. The wrapping keeps the link from melting until fire has spread to and through the doorway. The fire then burns away the wrapping; the link melts, the door closes, and supposedly the investigator is fooled. As is true of most arson devices, however, this device leaves evidence of its use: Pieces of the wrapping usually adhere to the linkage on either side of the fusible link, where they may be observed by a thorough investigator.

Fire Protection Systems

Automatic heat, smoke, and flame detection devices should be examined carefully—by an expert, if necessary—if they failed to transmit the alarm or if the alarm was delayed. The failure of a Halon, carbon dioxide, or other, more complex system to contain or extinguish a fire may indicate tampering or circumvention; the investigator should not hesitate to ask for the help and advice of the manufacturer of the system. A sprinkler system that fails to operate properly should, however, be examined by the fireground investigator.

The failure of a sprinkler system to operate at all may be due to closed water supply valves or broken piping. A closed water supply valve may also be damaged so that it cannot be opened; in either situation, the investigator must determine why the valve was closed. In some cases, an open supply valve has been damaged so that it could not be shut off once the sprinklers were activated; the object was to increase the water damage to contents.

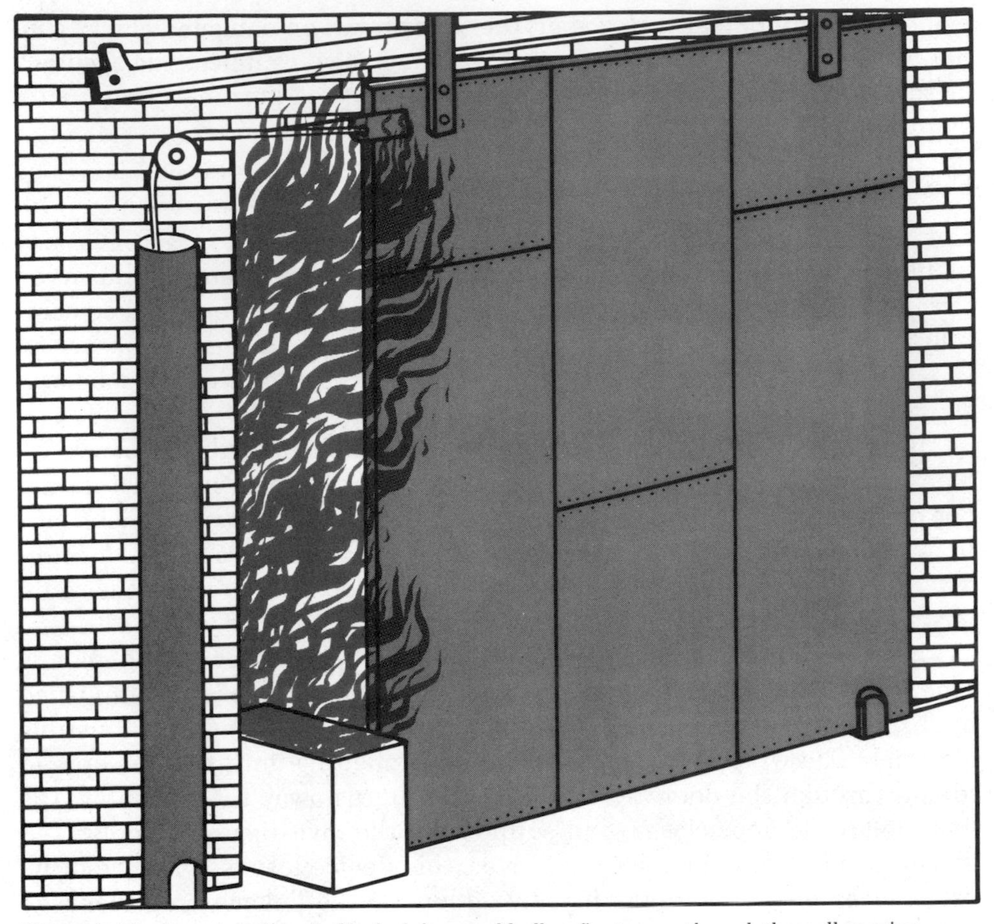

Figure 7-34. *An unintentionally blocked door could allow fire to pass through the wall opening.*

The pipes of a wet system may be damaged by opening a window near the piping in cold weather. This causes ice to form in the pipes, which evenutally ruptures them and releases water. Although this is not arson (because no fire is involved), it could be a prelude to arson; the sprinkler system might have to be taken out of service for several days while repairs are made.

The pipes of a dry system may be damaged in a number of ways; many of them are undetectable until the system operates improperly when the water is pumped into it.

Inadequate pressure in a sprinkler system may be due to a partially closed supply valve; to damaged piping that allows water to flow out of the system before it reaches the fire; or to the activation of an excessive number of sprinkler heads.

The activation of an excessive number of sprinkler heads may be accidental or deliberate. Normally, two heads should control a fire; no more than four or five should ever be operating at one time. However, a heavy coat of paint will delay the operation of the heads, allowing the fire to gain headway and generate heat over a wide area. When the heads finally are activated, an excessive number will open. Besides reducing the size of the spray pattern at any one head, this tends to increase the water damage.

Too many sprinkler heads may also open when fire flashes over an area. If the flashover cannot otherwise be explained, it may be the result of the use of an accelerant.

CASE HISTORY: Missing Contents

Firefighters responded to an alarm of fire in one apartment of an expensive apartment building. They forced entry into the apartment and extinguished the fire with some difficulty. The living room was most heavily involved, but the fire had extended to the kitchen and both bedrooms.

The firefighters and the fireground investigator observed much that indicated arson:

- Heavy black smoke, reddish flames and the smell of gasoline; the involvement of hydrocarbons was verified with a hydrocarbon detector
- Fire travel in two opposite directions from the probable point of origin, indicating the use of an accelerant
- Venetian blinds down and closed, probably to delay discovery
- Difficulty in extinguishment, and fire travel under the closed kitchen door
- Straight-line burn patterns along the living room rug, from the kitchen to the bedrooms

The apartment was declared a crime scene. The locked apartment doors, noted by attacking firefighters, and the absence of signs of illegal entry seemed to point to the occupant as the arsonist. He maintained that he had been out of the apartment all day, and had arrived only after the fire was extinguished.

Further investigation yielded two additional items of evidence: First, a neighbor mentioned during questioning that the occupant collected paintings; but the investigator could not find a painting or the remains of a painting in the apartment. Second, the staining on the living room and bedroom walls was uniform, indicating that if paintings ever had hung there, they were removed prior to the fire.

When confronted with this evidence, the occupant confessed to the crime. He had removed the paintings and set the fire, using gasoline as an accelerant; he had expected to be paid for the paintings by his insurance company, to use the money to pay off a debt, but to retain the paintings.

The fireground investigator, who was still at the scene when police arrested the occupant, reported the following:

Area of origin: Living room sofa
Heat source: Delay device consisting of a lit cigarette in an open matchbook
Reason: Deliberately set, with gasoline as accelerant
Category: Arson.

8

EVIDENCE

To this point, we have used the word *evidence* in two ways. Informally, evidence is any clue or indication as to what might have occurred before or during a fire. Almost anything observed by firefighters and investigators at the scene could be called evidence in this sense of the word. However, in the legal sense the word has a much narrower meaning: Evidence is something that may be presented in a court of law to establish facts. It must comply with very restrictive *rules of evidence* established by the courts. If it is not collected, documented, and handled properly, it may be ruled inadmissible by a judge.

This chapter is concerned with evidence in the legal sense. The discussions involve procedures rather than statutes and should apply in all 50 states. However, because the rules of evidence vary somewhat from state to state, the individual investigator must become familiar with the requirements of his own state. The fire marshal's office can be of help in this regard.

TYPES OF EVIDENCE

Evidence is classified in two ways—according to form and according to content.

Types by Form

The three forms of evidence are:

1. *Testimonial,* or spoken, evidence
2. *Documentary,* or written, evidence
3. *Physical,* or material, evidence.

The fire investigator is concerned with all three types, on the fireground and (if necessary) in court. Moreover, he must be concerned with all evidence that establishes the facts—evidence that supports innocence as well as evidence that indicates guilt.

Testimonial Evidence. Most of the evidence presented at a trial is testimonial evidence—the sworn statements of occupants, witnesses, police, investigators, and others involved in the investigation. Even when documentary or physical evidence is presented, testimonial evidence is used to explain or clarify it.

At the fireground, the investigator is concerned with statements that may later be repeated in court as testimony. They begin with firefighters' observations, include the statements of witnesses and occupants, and may also include the statements of technical experts (if an advanced investigation is required). Since the investigator is the first person to hear this evidence in an official capacity, it is important that he make an accurate record of its content. This is, of course, true for the other forms of evidence as well.

Documentary Evidence. This is written evidence that is important for its content—for what is written—rather than for the fact that it exists. As an example, suppose an investigator finds records indicating that a business has been losing money steadily; those records could be *documentary* evidence of a motive for arson. However, if the residue of a flammable liquid were found on the records, they would be *physical* evidence of the use of an accelerant.

An investigator may discover documents of various types during his interior examination. The documents should be left in place unless they may become damaged by water or by cleanup operations. To do otherwise is to violate the privacy of the owner. Even if arson is suspected, the investigator should attempt to preserve documentary evidence in place (where it was found) unless that is impossible or the police or fire marshals ask that it be removed.

The types of documentary evidence that may be discovered or obtained during a fireground investigation include:

- Business records, which, as noted, might indicate financial instability and, hence, a motive for arson
- Cancelled checks, which could establish the recent purchase of additional fire insurance or arson supplies
- Written statements of witnesses and others who may have information concerning the fire or the fire structure
- Written and signed confessions obtained during the questioning of witnesses
- Suicide notes found at the scene, and notes threatening occupants of the fire building
- Written reports of the results of laboratory tests.

Physical Evidence. Physical evidence is tangible material that has substance and dimension, and that provides information because of what it is. A can containing flammable liquid, a footprint, or a char pattern would be considered physical evidence.

Some types of physical evidence can be presented in court—a piece of smoke-stained glass or an arsonist's timing device, for example. However, it would be extremely difficult to transport wall burn patterns to a courtroom for use as evidence. In such situations, the court will allow the use of *documentary* physical evidence consisting of photographs, diagrams, and models that represent the actual physical evidence. A plaster cast of a tire track is an example of such documentary physical evidence.

Most of the evidence observed and gathered by the fireground investigator is physical evidence—either actual or documentary. The fact that documentary physical evidence is admissible in court lends great importance to the investigator's photos and diagrams.

Types of Content

Any of the three forms of evidence may be either direct or circumstantial. *Direct evidence* is evidence that, if true, immediately establishes what is to be proved. *Circumstantial evidence* is evidence of a condition or circumstance that is usually accompanied by what is to be proved. Circumstantial evidence thus implies, but does not by itself establish, what is to be proved.

As an example, suppose a witness at an arson trial testifies as follows: "I saw Smith walk into his house carrying a five-gallon can. The way he carried it, it looked very heavy. About three or four minutes later, he ran out of the house, carrying the can easily, and drove away in his car. Right after that, I saw flames through the windows." This is circumstantial evidence; the testimony implies that Smith emptied and ignited the contents of the can in the house, and that this was the cause of the fire. However, the testimony does not prove that he did so.

Now suppose another witness testifies that "Through the windows, I saw Smith carry a can into his living room, pour a liquid all over the furniture and the rug, and light it with a match. Flames shot up all over the place, and he ran out of the room." This is direct evidence; if it is true, it establishes, without supporting evidence, the fact that Smith started an arson fire.

Most arson convictions are based on circumstantial evidence of all three forms, each piece of evidence reinforcing the others to provide a reasonable case against the arsonist. Occasionally, an investigator may find direct evidence pointing to an arsonist; most often, though, he will be concerned with the observation and preservation of circumstantial evidence discovered on the fireground.

DETAILED SEARCH FOR EVIDENCE

When a fire is of suspicious origin or is obviously an arson fire, the involved structure is searched for evidence of the crime and the criminal. The search centers on the point or area of origin. The searchers seek such items as arson devices, containers and residues that may be related to flammable liquids, remnants of a missing heat source, signs of robbery or vandalism, and burn patterns that may be hidden under debris. In some cases, partially consumed documentary evidence indicating the motive for the crime has been found among the debris.

Small but important items of evidence might easily be missed in an unplanned or hurried search. The search area should be gone over systematically and thoroughly. The searcher or searchers should work through the area slowly, *observing* what they see. Once the search area has been delineated, it should be searched completely. An offhand decision to skip even a small part of the search area may negate all the effort expended in the search.

As usual, the searchers should wear protective gear and carry handlights.

Search Patterns

In most cases, the search area will be the room in which the fire originated. If the room is not very large, a pattern similar to that used in search-and-rescue operations may be employed. Begin at one corner of the room, about 2 feet from the wall closest to the point of origin. Move parallel to that wall, across the length of the room, examining a strip of the floor about 4 feet wide. Turn at the far end of the room, move about 4 feet further from the wall, and again cross the room while examining a 4-foot strip of the floor. Continue this pattern until the entire room has been searched.

If the room is large, it should be divided into smaller areas before the search. Then each smaller area may be searched individually by implementing this pattern. The equipment kit suggested in Chapter 6 includes rope, markers, and pennants, which may be used to divide the area for the search. This equipment may also be used to cordon off a larger area (consisting, perhaps, of several rooms) while it is being searched. The markers and pennants serve both to outline the search area and to warn firefighters that a search is being conducted.

Before searching a particularly large area, it is a good idea to divide the area with a grid made of lightweight rope or markers. For example, an area of several square yards may be divided into areas of a few square feet each. Small cans painted orange or yellow for visibility make excellent markers. They also give warning that a search is in progress. The markers are placed so that they form a grid indicating the corners of small square areas. Each of these areas is searched along one side, then back through the center, and finally along the opposite side.

Sifting Debris

Many times a search for small pieces of evidence requires that the debris be sifted through a mesh. A fairly coarse mesh will serve for most situations for locating such items as matches, melted metal from wires, pieces of light bulbs, and bits of cloth.

The debris from a particular grid area should be sifted onto a sheet of newspaper or a piece of canvas, and then discarded outside the search area. This avoids confusion as to which debris was already examined and which wasn't.

CASE HISTORY: Contamination of Evidence

A fire involving every room in one apartment of a three-story apartment house was extinguished only after the fire had rekindled in several rooms. The circumstances were so obviously suspicious that the shift commander called for an arson investigator without delay.

During firefighting operations, the occupant of the involved apartment had twice asked the shift commander for permission to enter the building to remove his valuables. The shift commander had refused both times, but said he would have the valuables protected. During both conversations, the shift commander could smell the odor of gasoline on the occupant's clothing. When the occupant realized that the shift commander would not let him enter the building, he approached another officer from whom he managed to obtain permission to enter the structure.

The investigator found enough evidence to prove arson. Based on the shift commander's information, he requested that police take the occupant into custody and confiscate the occupant's clothing for laboratory analysis. The occupant's shoes were wrapped in the clothing, and the clothes were packaged and sent to a laboratory. When the analysis confirmed the presence of gasoline in the clothing, the occupant was charged with arson and brought to trial.

The defense contended that the gasoline on the defendant's clothing had rubbed off the shoes that were packaged with the clothing. Further, the defense maintained that the defendant's shoes picked up the gasoline when he entered the apartment after the fire was extinguished. Since there was no evidence (other than the clothing) to prove that the occupant handled gasoline prior to the fire, he was acquitted.

The message of this case is clear: Keep unauthorized persons off the fireground until the investigation is completed. The message also has two postscripts:

1. Occupants are unauthorized persons, as much as anyone else.
2. Every officer and crew member should be aware of the message.

COLLECTING AND PRESERVING EVIDENCE

Evidence that is discovered by fireground investigators should be preserved in place, where it was found, if at all possible. The evidence should not be moved until arson investigators have had a chance to examine it. Fire department personnel may retain possession of a premises, and deny access to owners or occupants, during an investigation. However, this legal right is subject to the requirement that the premises be returned within a reasonable time. Although there is no way to measure a "reasonable time," a delay of more than two or three hours after completion of the fireground investigation could be considered unreasonable. If arson investigators have not arrived within these few hours, the fireground investigators must remove the evidence from the premises.

Fire investigators have the right to confiscate physical evidence that they believe is related to a suspected crime. They must handle and package it properly to avoid contamination; provide the owner with a receipt showing each item removed from the premises; and ensure that the evidence is stored properly (the fire department should have a room, with a strong lock, set aside for this purpose).

In collecting evidence that may be used in court, the investigator must make sure it is

- Witnessed
- Photographed in place
- Diagrammed
- Marked
- Packaged properly
- Sealed in its container
- Identified and recorded
- Properly stored.

These steps ensure that the origin and integrity of the evidence are verified and that the evidence is not contaminated. They provide the greatest assurance of acceptability and evidential strength in a courtroom.

Witnesses

While the evidence is still in place, the investigator should ask at least one witness to observe its existence and location. If possible, he should obtain two witnesses for increased validity. In most cases, the witnesses will be firefighters or police officers, since unauthorized persons should be barred from the fireground. If emergency personnel have left the scene, the evidence should be witnesses by individuals who are not directly involved with the fire.

The names and affiliations (or addresses) of the witnesses should be recorded as detailed below.

Photographs

Each piece of evidence must be photographed from several angles, while it is still in place. The photos will serve to reinforce the evidence, show the conditions under which the evidence was found, and serve as demonstrative physical evidence in the event that the actual evidence cannot be brought into court.

The pertinent information concerning each photograph should be recorded (see Chapter 6).

Diagrams

The investigator should draw a diagram showing the location of the evidence when it was discovered. As usual, the diagram must not be drawn to scale. However, the distance of the evidence from fixed objects (a window, doorway, radiator, etc.) should be measured as accurately as possible and noted on the diagram. (The distance of a piece of evidence from a portable object, such as a chair, becomes meaningless if that object is moved.)

The investigator should note, on the diagram and in a notebook, what the evidence is, where and when it was found, all identifying marks such as dents and stains, and a brief explanation of the significance of the piece of evidence.

Preparation for Removal

At this point, the evidence may be handled and moved. The evidence should not, however, be touched with bare hands, which would leave fingerprints on it. (This obviously does not apply to demonstrative physical evidence, which may be handled freely.) Small pieces of evidence may be lifted with tweezers; larger pieces, with tongs. If necessary, the investigator should don gloves before handling more bulky pieces.

Samples of woodwork, flooring, rugs, or furniture that have to be cut out for laboratory analysis should be treated in the same way as other evidence. Where possible, comparison samples should be obtained, as described below.

Containers. A number of different types of containers should be available in the investigator's kit, for the packaging and preservation of evidence. These containers are listed in Chapter 6. In most instances, the evidence itself will dictate the type of container to be used. In general, liquids and items of evidence that are soaked with liquids require containers that will hold the liquid but not contaminate it, and that are airtight to prevent evaporation. Small items, such as matches and cigarette butts, may be placed in containers. Cans used to hold evidence should be unlined; the linings may contaminate some types of evidence.

Only one item of evidence may be packaged in each container—even if the items are exactly alike. The reason is, again, to guard against the contamination of one piece of evidence by another.

Comparison Samples. When a flammable liquid is poured or splashed about an area, it penetrates into those parts of furnishings and building features that it contacts. It may, for example, saturate one part of a rug or one section of baseboard, but not touch the remainder. When the liquid is ignited, it causes deep burning on the saturated parts, but the remainder burns normally (or not at all). In many instances of arson, accelerants are thrown onto clothing in a closet; the clothing then exhibits this same type of telltale burn pattern. A sample of the heavily burned material should be taken for analysis and as evidence of the use of an accelerant; a similar sample should also be taken from the lightly burned (or unburned) section, for comparison.

The burn patterns should indicate clearly where the flammable liquid penetrated and where it did not. When fabric samples are to be cut from a rug, drapes, clothing, or upholstered furniture, a piece should first be cut from the deeply burned area; then a similar piece should be cut from an area as far from the first as possible. The same procedure should be used in cutting samples from wood, but both samples should be cut from the same board. This will ensure that the wood itself is the same as in both samples.

The samples should be marked as discussed in the next section and packaged in separate but identical containers. (Identical containers are used to ensure that, if the containers affect the samples at all, both will be affected in the same way.) Airtight containers should be used to prevent the evaporation of flammable-liquid residues. The containers should be sealed and the evidence recorded as detailed below.

The samples should be sent to a laboratory for analysis as soon as possible, to minimize evaporation. If a delay is unavoidable, the containers may be refrigerated to retard evaporation.

Investigator's Mark

The investigator's individual mark (usually his initials) should be placed on each piece of evidence, if it is possible to do so. The mark may be cut, scribed, or filed onto solid objects; cloth is usually marked with waterproof ink. The mark ensures proper identification in court, identifies the item in the event it is confused with other investigators' evidence, and serves as evidence that the investigator actually collected the item and is qualified to discuss its location and condition when found.

Every item of evidence should be marked, whether it is to be packaged or not. Many jurisdictions require that specific rules be followed when evidence is marked. Either the police department or the fire marshal's office should be able to provide the details.

Packaging

Physical evidence may or may not require packaging, depending on its material and size and the amount of protection it needs. Very fragile materials, such as burned paper and cloth, should be preserved in place if at all possible, until arson investigators arrive. Large items, such as television sets and electric motors, are too bulky to package. Other items may not require protection. For example, an electric clock may be important only because its hands indicate the time at which it stopped. The hands may be photographed and then sealed in position with tape, but the clock need not be packaged.

In the next several paragraphs, we discuss methods of packaging the physical evidence most often found after a structure fire. Whether a piece of evidence is packaged or not, all the other steps in the evidence collection procedure must be performed.

Paper. As noted, burned paper is very fragile. The best way to preserve it is to protect it in place. A cardboard box, a dishpan, or a bucket may be placed over the paper—open end down—to form a temporary protective cover. The cover should be lowered into place slowly and carefully, so as not to touch the paper or create an air current that might disturb it. The cover should then be clearly identified, so that its purpose is immediately understood by anyone who sees it. The warning

Evidence being protected, DO NOT TOUCH

printed on the cover with a felt-tip pen should be sufficient.

Another method requires several pieces of wood and a sheet of transparent plastic. Pieces of wood are placed around the evidence. Then the plastic is placed over the wood carefully and stretched and held in place with other available objects. The use of the transparent plastic allows the investigator to view the evidence while the cover is being placed and, later, when it is removed. A salvage cover may also be used as a protective tent over fragile evidence; care must be exercised so as not to disturb the evidence when the cover is positioned and removed. With either the transparent sheet or the salvage cover, a warning sign should be displayed prominently, to ensure that the evidence is not destroyed.

When a sheet of burned paper must be removed from the fire building, it may be packaged between two pieces of plate glass. Figure 8-1 shows a burned document that is to be packaged. The paper is gently pushed onto one piece of glass (Figure 8-2). The second piece of glass is then carefully placed on top of the paper. The glass is then sealed around all edges with tape and wax (Figure 8-3) and tagged as evidence, as detailed below. The glass supports the paper so that it will not fall apart, and allows the evidence to be examined while it is still "packaged."

Figure 8-1. *Burned paper is fragile and must be protected.*

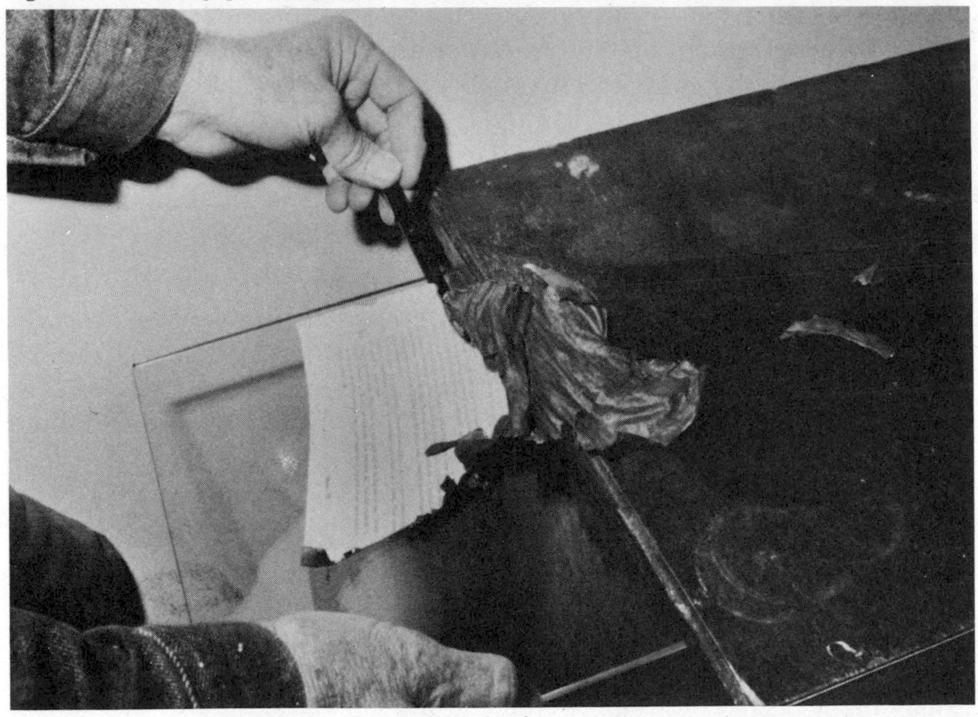

Figure 8-2. *Carefully push the burnt paper onto the glass.*

Burned paper may also be packaged in dinner plates. The paper is carefully pushed onto one plate and the second plate is placed on top of the first. The plates are taped, sealed, and tagged. Because the plates are opaque, they must be taken apart when the evidence is viewed; the evidence could be damaged in the process.

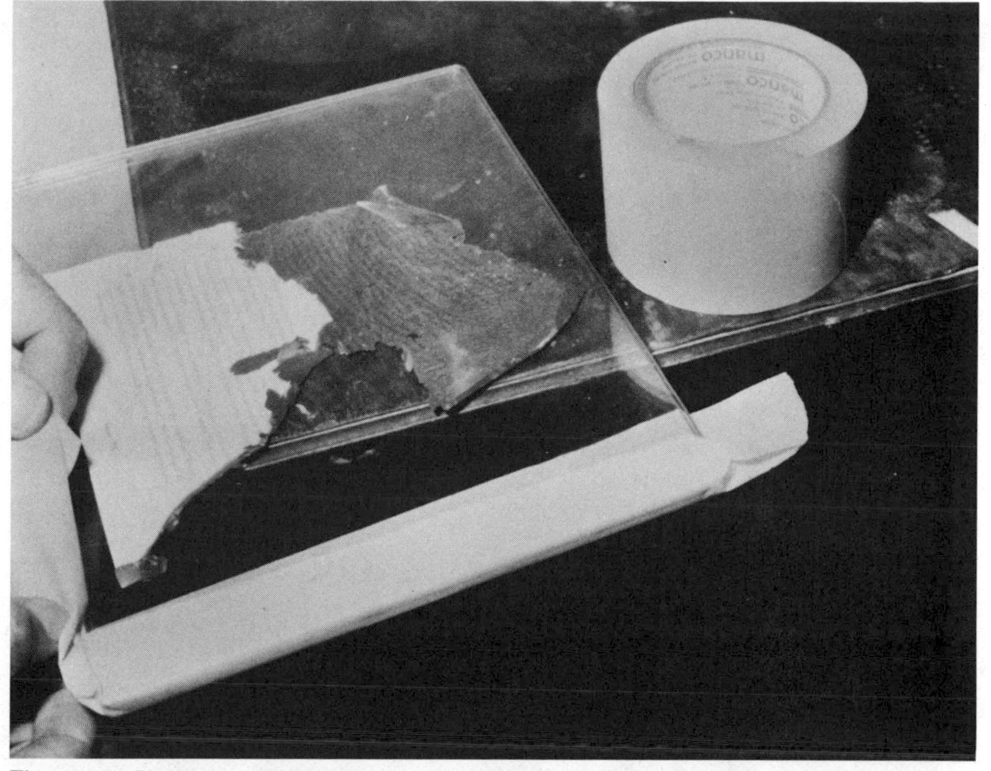

Figure 8-3. *Place a second piece of glass over the paper and seal it with tape.*

Fabrics. Burned cloth is as fragile as burned paper. Thin fabrics that are suspected to contain writing should be packaged in the same way as burned paper. A fragile piece of cloth may also be lifted with tweezers or tongs and sandwiched between two pieces of cardboard, taped, and sealed. Bulkier fabrics may be handled normally but carefully. Any fabric that is suspected to contain a flammable liquid or its residue should be packaged in an airtight container.

Figure 8-4 shows a rug from which a sample is to be cut for analysis. The sample is cut out with a knife and lifted with a pair of kitchen tongs (Figure 8-5). The sample is placed in an unlined paint can (Figure 8-6) and the cover is placed on tightly. Tape is wrapped around the container, with the ends overlapping at the top. The tape is then sealed with wax and the container is tagged as evidence. The container must be handled with reasonable caution to prevent damage.

If a comparison sample of the rug were required, it would be cut, handled, and packaged in exactly the same way (Figure 8-7).

Wood. Wooden objects that are to be removed from the fire building as evidence may be packaged in almost any type of container. If the object is large and reasonably sturdy, it may not require packaging at all. Wooden objects or samples that are suspected to contain flammable liquids or residues must, as usual, be packaged in airtight containers.

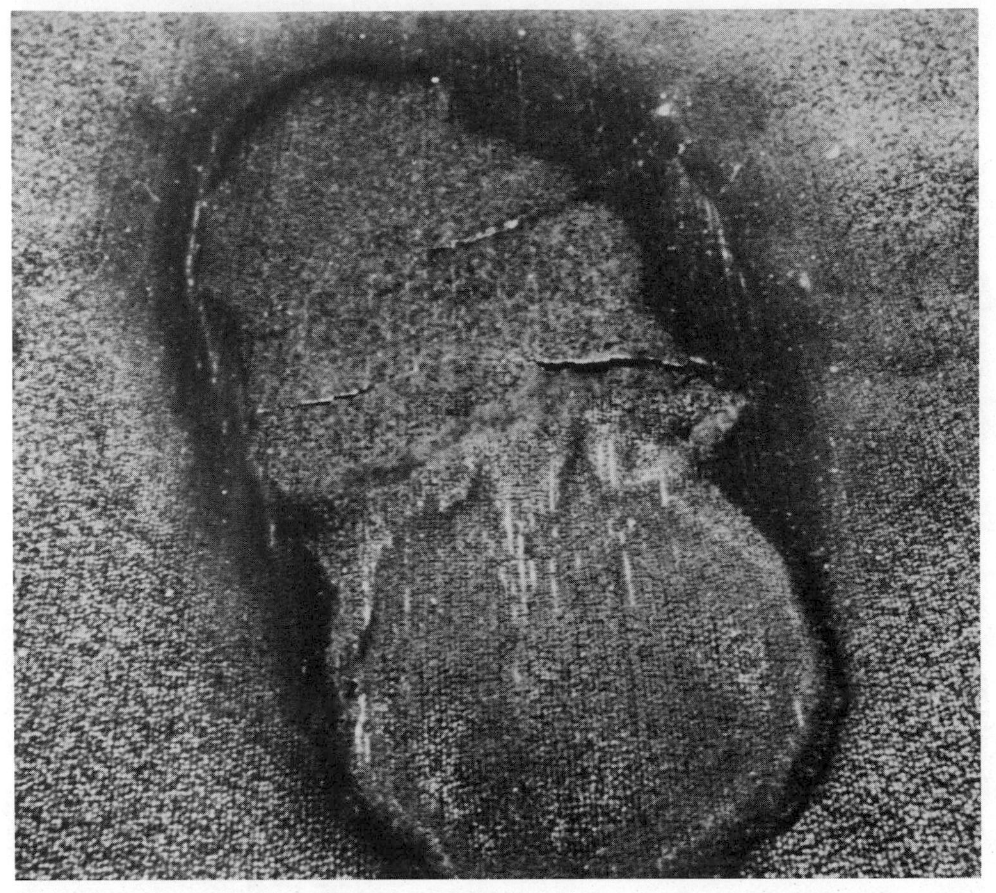

Figure 8-4. *Remove a section of burned carpet for an analysis.*

Figure 8-8 shows a wooden baseboard from which a sample is to be cut for analysis. The baseboard is close to the point of origin of a fire that seems to have involved a flammable liquid. A small keyhole saw is used to cut out a section of the baseboard (Figure 8-9). The sample is lifted with kitchen tongs (Figure 8-10) and placed in a new, unlined paint can. The cover is replaced tightly and tape is wrapped around the can, with its ends overlapping (Figure 8-11a). The container is then sealed and its contents recorded. A comparison sample should be cut from the same baseboard, at some distance from the point of origin, and packaged and sealed in the same manner (Figure 8-11b).

Glass. The way in which glass should be packaged depends on why it is being preserved. For example, smoke-stained window glass must be packaged so that the stains are not smudged, whereas a bottle that may have held an accelerant must be placed in an airtight container. All glass should be handled with tongs, if possible; broken glass should be grasped by the edges so that surfaces are not disturbed.

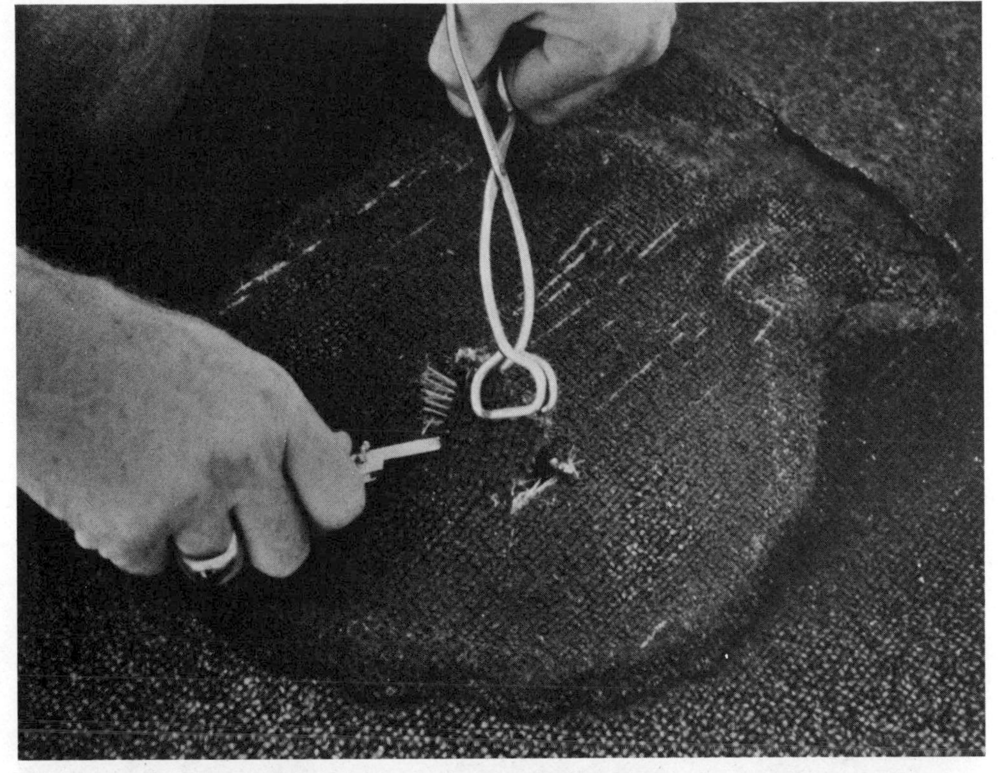

Figure 8-5. Cut a piece of burned carpet and remove it with tongs.

Figure 8-6. Place the carpet sample in a can and seal the cover.

Pieces of smoke-stained window glass may be placed in a cardboard carton with slots cut to hold the glass in place. They may also be wedged

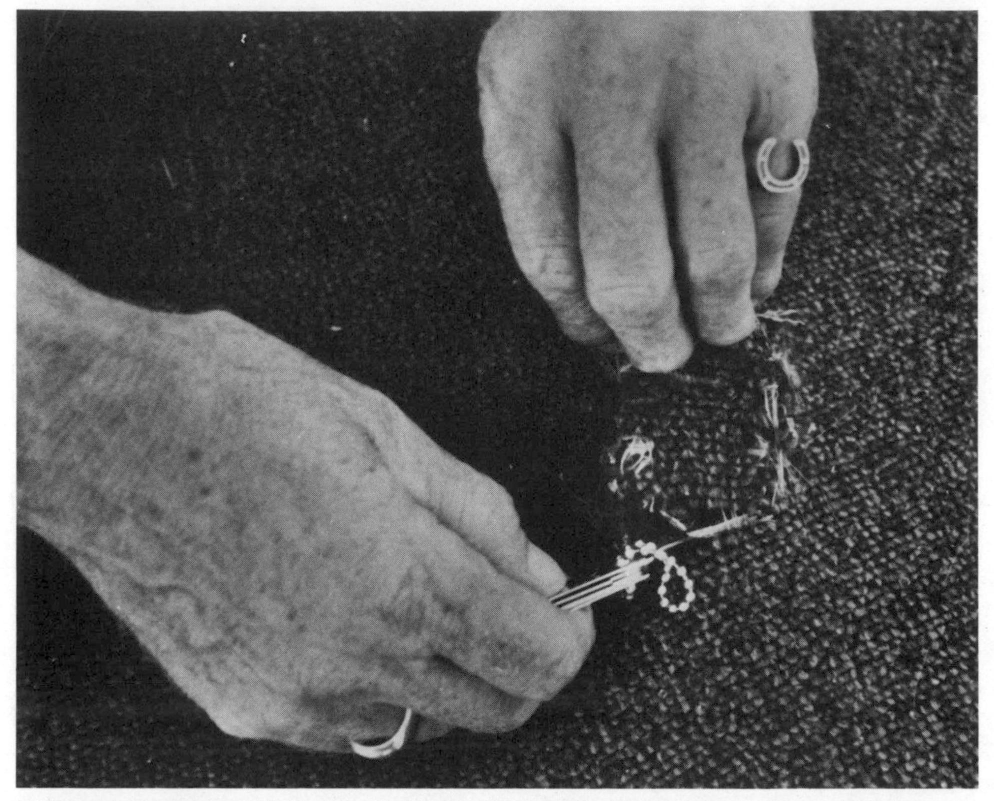

Figure 8-7. *For comparison, package an uncontaminated piece of carpet.*

Figure 8-8. *Take a sample from burned baseboard molding.*

Figure 8-9. *A keyhole saw can be used to remove a burned section of baseboard molding.*

Figure 8-10. *Place sample into an unlined can with tongs.*

into styrofoam cups—one piece per cup—if they are small enough to fit (Figure 8-12). Or the glass may be placed on a piece of cardboard and

Figure 8-11a. *Seal the container with tape.*

Figure 8-11b. *A comparison sample of an uncontaminated baseboard should be analyzed.*

Figure 8-12. *Styrofoam cups make excellent containers for small pieces of glass.*

covered with a second piece of cardboard; the two pieces of cardboard should be taped together tightly and then taped all around, sealed, and recorded as evidence. The tape should not be applied directly to the glass, since it could lift stains or prints off the surface of the glass when it is removed.

Glass that is collected as evidence because it shows signs of melting can be wrapped in newspaper and placed in a carton. The wrapped glass should be surrounded with crumpled paper to keep it from moving within the carton.

Glass that carries fingerprints or bloodstains should be handled very carefully; its surfaces should never be touched. It may be placed in a container and should then be sealed.

A glass container that is suspected to have held an accelerant should first be sealed in an unlined can. Then the can should be placed in a box and cushioned with rags or crumpled paper. (The cushioning material should not be material that was involved in the fire.) Pieces of glass from a container that may have held accelerant may be packaged in the same way. Each piece should be sealed in a separate unlined can but they may all then be placed in the same box.

As usual, the outer container should be taped, sealed, and recorded as evidence.

Electrical Items. Electric wires and components are quite rugged and may be packaged in heavy manila envelopes or in boxes. When an envelope is used, the evidence should be placed inside, and the flap sealed; tape should be wrapped completely around the top of the envelope to cover the flap.

When a box is used, tape should be wrapped completely around it, with the ends overlapping at the top. If brittle wire, or wire with brittle insulation, is packaged in this way, the container should be stuffed with crumpled paper as added protection against damage.

Small pellets or balls of melted copper may be packaged in plastic medicine bottles or other small containers.

Gas Valves, Pipes, and Meters. Gas-system components are fairly rugged and need not be packaged unless fingerprints or other markings must be protected. They should, however, be properly recorded and tagged as evidence.

If possible, the components that are needed as evidence should be dismantled from the system as a single unit. That is, they should not be disassembled into individual components unless they cannot be removed any other way. When the components must be disassembled, each of them should be identified on photographs and sketches of the whole assembly, as it was found by the investigator.

Tools. Crowbars, screwdrivers, tire irons, and other tools that may have been used for forced entry must be handled and packaged carefully. Fingerprints, minute particles of paint or wood, blood, and residues can be discovered during laboratory analysis and matched with the crime or a suspect.

Two sheets of cardboard may be used to package a tool. Slots are cut in one sheet of cardboard to accept the tool. The tool is slid into the slots with a pair of tongs (Figure 8-13). The second sheet is then placed over the tool and the edges of the two sheets of cardboard are taped on all four sides (Figure 8-14). Tape is also wrapped completely around the package, with the ends of the tape overlapping.

A tool may also be packaged in a cardboard box. Again, the tool should be lifted and placed in the box with tongs. Tape should be wrapped around the box in two perpendicular directions, with all four ends joining at the top of the box.

If a tool found on the fireground seems to match pry marks on a wooden window sill or door frame, a small piece of the wood, including the marks, should be cut out and packaged as evidence. The marked wood should be sent to the laboratory with the tool for a detailed comparison.

Small Items. Buttons, matches, and other small items of evidence may be packaged in letter-size envelopes or pill boxes provided there is no concern about flammable liquid vapors. Tweezers should be used to lift and handle these items to avoid contamination (Figure 8-15). When a pill box is used it should be taped all around, with the ends of the tape overlapped at the top. An envelope should be sealed first with its flap and then with tape running around the center and covering the flap.

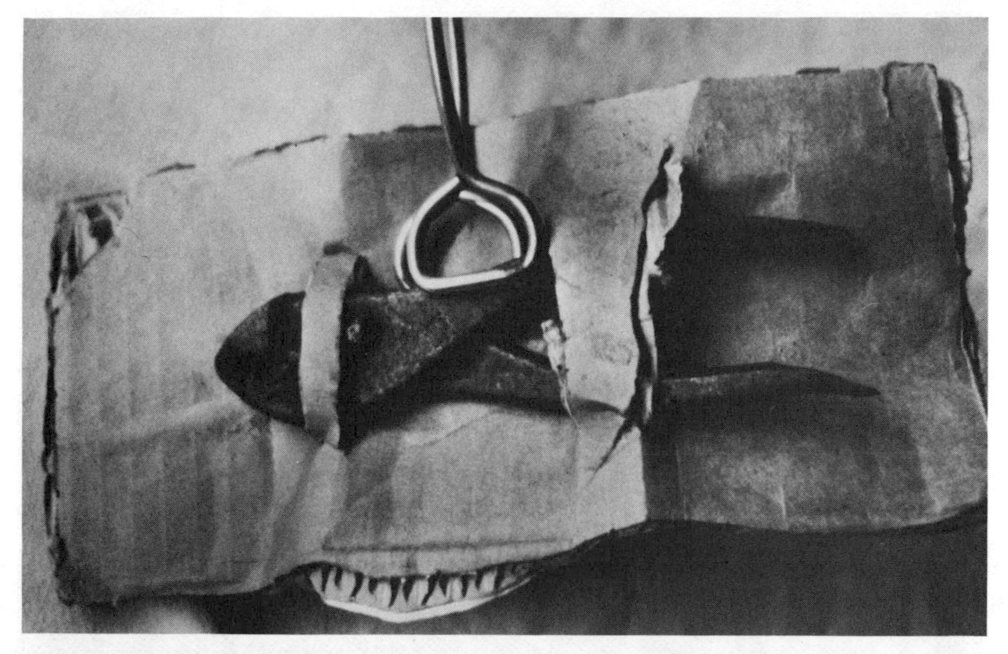

Figure 8-13. A receptacle for a bulky tool.

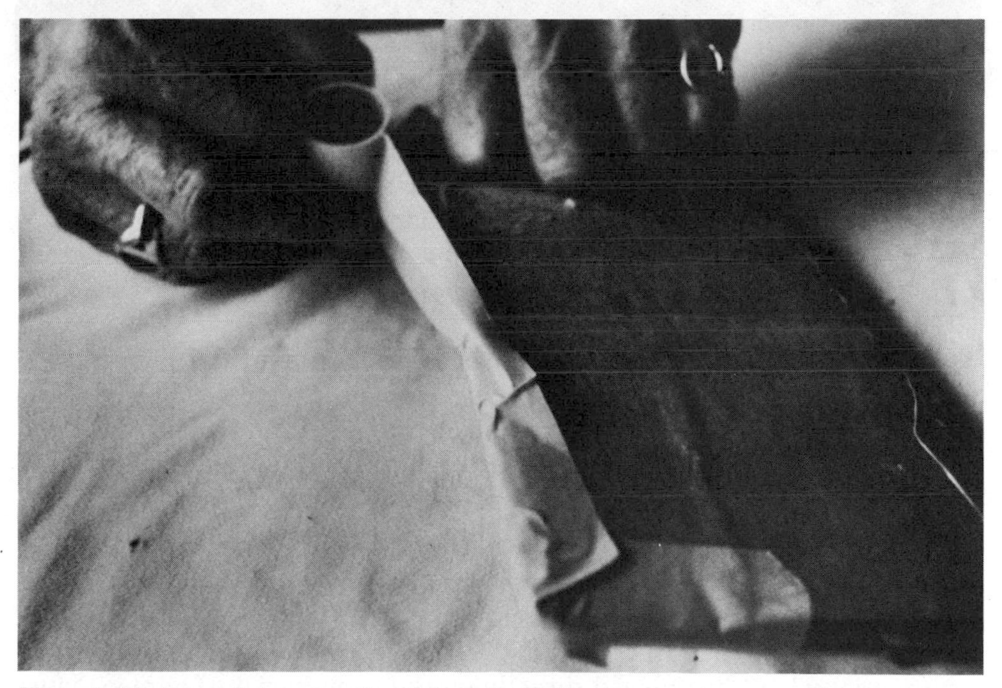

Figure 8-14. A second piece of cardboard forms the final package.

Sealing

In each of the packaging methods described in the preceding section, tape was wrapped around the container with the ends of the tape overlapping

Figure 8-15. *Use tweezers to handle fragile materials.*

or joining. This is actually the first step in sealing the evidence to protect its integrity. Masking tape may be used for this purpose. It is best if the tape makes several loops around the container. The tape should cross over any open seams on a box, the flap of an envelope, or the lid of a jar or paint can. The tape must form a single continuous loop or, if two loops are used all four ends should meet at one point.

Once the tape is in place, sealing wax is melted onto the joined ends. Figure 8-16 shows this step being performed on the paint can. Then, while the wax is still soft, the investigator should dip one finger in cold water and press a fingerprint into the wax (Figure 8-17). This squeezes the wax into the tape joint and provides positive identification of the packager.

Identification and Recording

After the container is sealed, its content must be identified. Quite a bit of information is required; it should be printed carefully and legibly. The information should be placed directly on the container if possible (Figure 8-18). Masking tape or a label may be attached to a paint can or a jar, and the information printed on that (Figure 8-19). Small containers and the packaging suggested for burned paper require tags containing the identifying information; the tags should be securely fastened to the containers. If a tag is lost, the evidence will probably not be admitted in court.

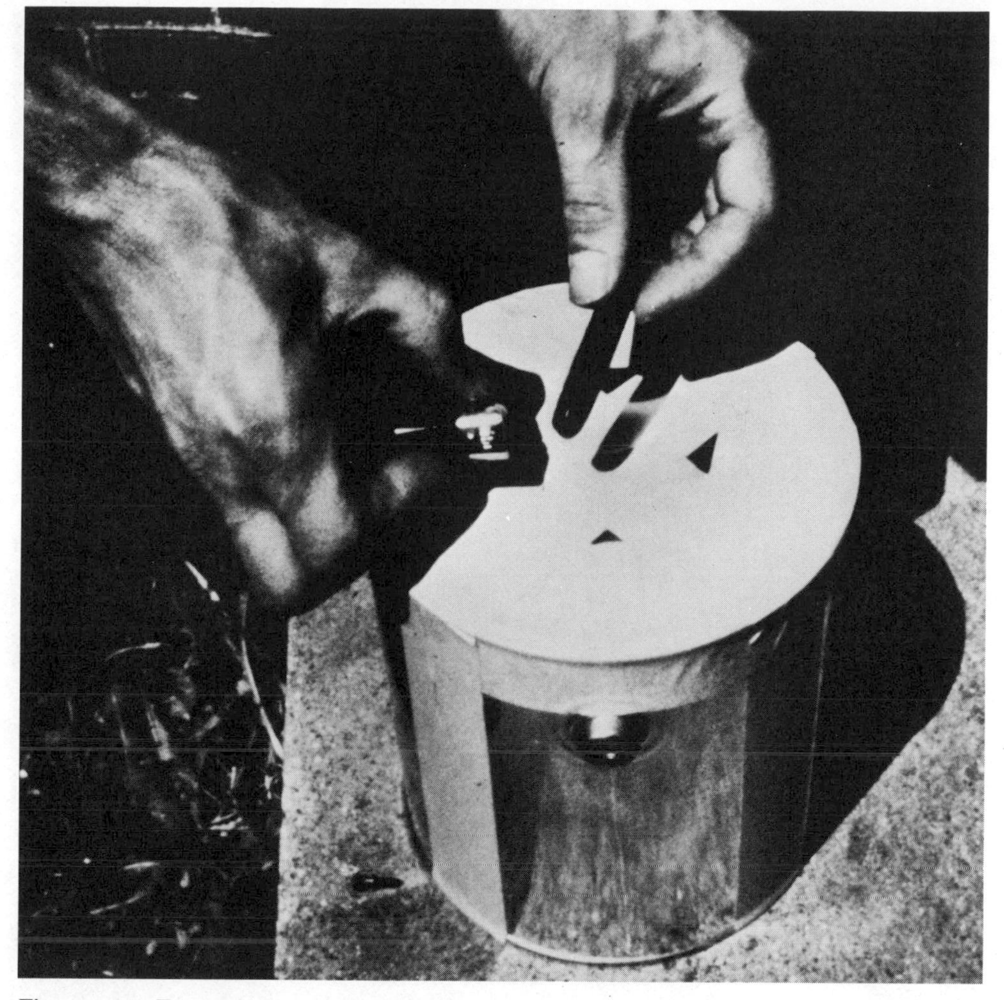

Figure 8-16. *To complete a sealing procedure, melt wax onto joined ends.*

The identifying information should include the following items:

1. *An exact description of the evidence.* The description should be concise and accurate; if the evidence is a comparison sample, that fact should be noted (see also item 7 below).
2. *The address of the structure from which the evidence was taken.* For a single-family dwelling, the street address is sufficient. For a multiple dwelling, office building, or hotel, the address must include the apartment, office, or room number.
3. *The exact fireground location at which the evidence was found.* Measurements from stationary features should be included, whether the evidence was found inside or outside the building. Compass directions should be used as necessary to describe the location. Sketches may also be used.
4. *The day, date, and time of day at which the evidence was found.*

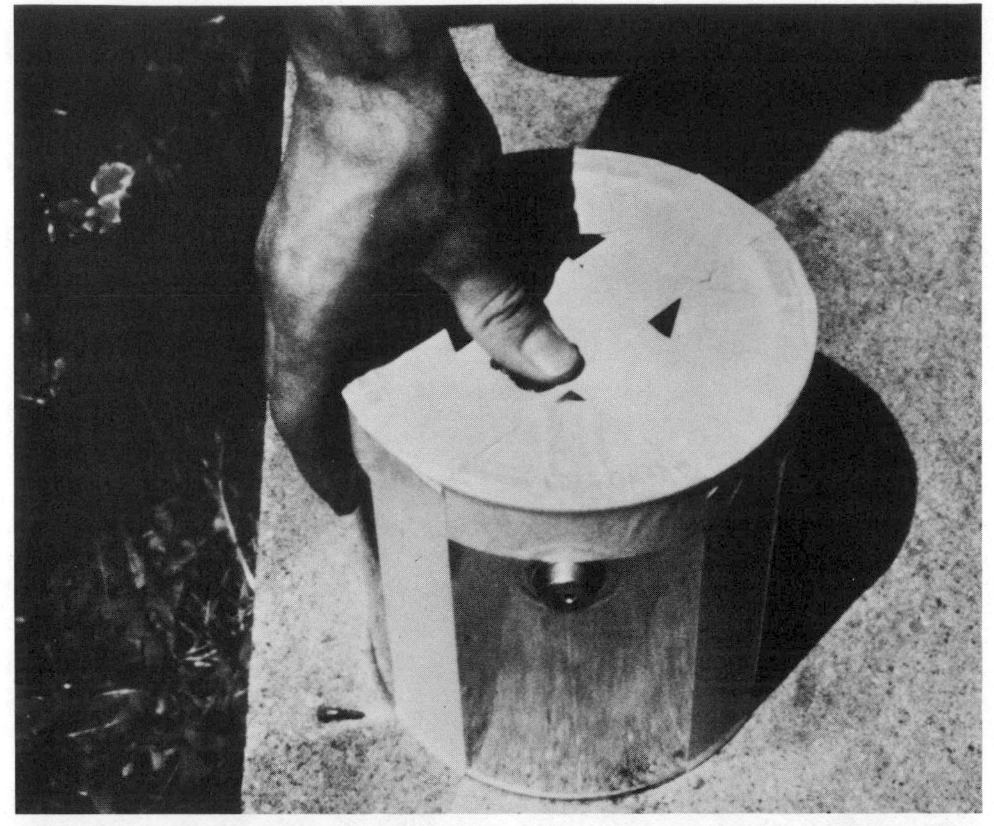

Figure 8-17. *Pressing fingerprint into the wax.*

Figure 8-18. *Write the information directly on the container if possible.*

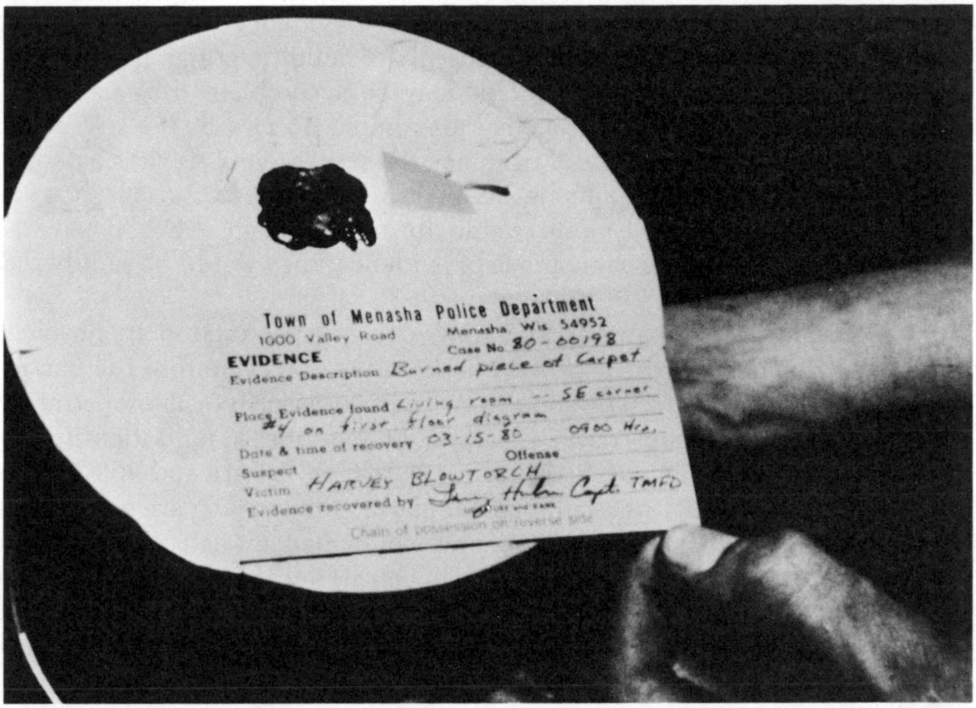

Figure 8-19. *Identify contents of container and securely fasten tag or label.*

5. *The fire incident number (alarm number).* This number should be obtained from the dispatcher.
6. *The evidence number assigned by the investigator.* Each piece of evidence should be assigned a number as it is collected. The most practical numbering method makes use of the alarm number, followed by a number indicating when, in sequence, each piece of evidence was collected. Thus, the second piece of evidence collected at alarm number 903 would be given the evidence number 903-2.
7. *The comparison designation (if applicable).* When comparison samples are collected, they are both given the same evidence number. The suspect evidence is given the additional designation A, and the comparison evidence the designation B. (This is standard throughout the country.)
8. *The investigator's mark.* This should be as close as possible a duplicate of the mark placed on the evidence itself.
9. *The investigator's name and fire department.*
10. *Witnesses' names and affiliations.* For firefighters or police officers, the full name, badge number, and fire company or precinct should be noted; for civilians, the full name, address, and telephone number.

The investigator must record each piece of evidence in his notebook, *duplicating exactly the wording that appears on the container.* In addition, each photograph and diagram pertaining to the evidence must be recorded in the notebook. The numbering of the photos and diagrams must correspond to the evidence numbers. This may be accomplished by using the prefixes P (for photo) and D (for diagram) and, if necessary, the suffixes a, b, c, and so on. Thus, if a piece of evidence is numbered 903-2, then a photograph of that evidence would be numbered 903-P2. A diagram showing where that evidence was found would be numbered 903-D2. If several photos were taken of the same piece of evidence, they would be numbered 903-P2a, 903-P2b, etc.

Demonstrative photographs and diagrams are not related to physical evidence. However, they should be numbered to correspond to the particular subject. The prefix DP should be used to identify a demonstrative photo, and DD a demonstrative diagram. The number should identify the subject. Thus, according to this system, photos 903-DP-1a and 903-DP-1b and diagram 903-DD-1 were all made at alarm number 903; are all demonstrative evidence (no corresponding physical evidence was collected); and all concern the same item (perhaps the low burn) because they all carry the number 1.

This numbering system may seem overcomplicated at first, but it is easy to use after a little experience, and it does a good job of organizing evidence. Its worth becomes very evident when an investigator who uses it is called to court several months after an investigation is completed.

Storage

Evidence collected at the fireground should be stored in a secure room at the fire station, set aside for that purpose. The door should be equipped with a strong lock, and keys should be issued only to fire investigators and the chief. Evidence must not be removed from the room without the authorization of the chief or an investigator.

THE CHAIN OF CUSTODY FOR EVIDENCE

The investigator who collects evidence at the fireground is responsible for that evidence—it is in his custody. When custody of the evidence is transferred to some other authorized person, that person must unpack and examine it to ensure that it is as described, repackage it, seal the container, record its acceptance, and provide a signed receipt to the original custodian. The original custodian must record and file the following information with other data pertaining to the incident:

- A description of the evidence that was transferred, including the evidence number (this information is taken from the original identification of the evidence)
- The name of the person releasing the evidence

- The name of the person receiving the evidence, including his badge number and affiliation or address and telephone number
- The date and time of the transfer
- The signed receipt for the evidence.

These records establish the chain of custody, ensuring that only authorized persons have had access to the evidence. They also relieve the original custodian of responsibility for the transferred evidence. If the transfer is not fully and accurately recorded, the court might rule the evidence inadmissible.

In most instances, the fireground investigator is involved in only one transfer of evidence—to arson investigators in an arson case or to the court if the fire results in a civil lawsuit. The investigator may not see the evidence again until he is called as a witness. It is for this reason that the investigator's notes are so important; they contain the only references that he can use to refresh his mind in the months (or years) between an investigation and the resulting trial.

9

INVESTIGATING THE FIRE: PEOPLE AND RECORDS

A fireground investigation is not complete until the investigator has questioned the people who are most directly concerned with the fire. These include emergency personnel, the owners and/or occupants of the fire structure, and persons who may have witnessed some aspect of the fire. Depending on the situation and the preference of the investigator, these people may be questioned before or after the fireground is examined. (Questioning and fireground examination may be performed at the same time fires are investigated by a team.) Their information may support and reinforce evidence discovered during the examination of the fire structure, or it may reveal entirely new information. If their information contradicts what is found in the building, then the investigator must resolve the contradiction.

Other people may be able to contribute background information to the investigation. This would include information concerning the occupants' normal routine and, perhaps, recent events of an unusual nature. Such background information (and some that is more directly related to the fire) may also be found in written records of various types.

This chapter is concerned with the gathering of information from these two sources—people and records.

PEOPLE WITH INFORMATION

As implied above, people may contribute two types of information to the investigation. The first, which may be called *direct* information, deals with the fire itself or with events that took place during or immediately prior to the fire. The second, *indirect* information, is concerned more with the general circumstances surrounding the fire; it may be obtained from people who did not actually witness the fire.

Direct Information

The people who are sources of direct information include emergency personnel, occupants of the fire building, and persons who witnessed part or all of the fire. These people are usually available at the fireground. They should be interviewed as soon as conditions allow, while the information is still fresh in their minds.

Firefighters. The numerous types of information that firefighters can provide to the investigator are discussed at length in Chapter 5. Firefighters' observations are mentioned again here to emphasize (1) their importance and (2) the fireground investigator's responsibility for "collecting" this evidence.

Police Officers. Police personnel sometimes arrive at the fireground before the first fire companies. They may then be able to provide useful information concerning the earliest stages of the fire. Moreover, police officers are trained observers of *people;* those engaged in crowd control often can detect unusual behavior in individuals watching the fire, or can recognize a face that seems to turn up at every fire in the district.

Police personnel, as well as firefighters, will do a better job of observing and providing information if they know what information is needed. It is up to the investigator and his superiors to make sure they do know.

Occupants. Occupants of the fire building who were near the point of origin may be able to describe what occurred. Or they may only be able to indicate where the fire originated, or where they first discovered smoke or flames. All such information should be carefully noted. In particular, the investigator should determine whether the statements of each occupant tend to confirm or contradict the results of his examination of the fireground.

Occupants who were not in the structure at the time of the fire may also contribute important information: the positions of contents before the fire; types and locations of unusual contents; difficulties they might have experienced with electrical appliances or circuits; unusual events that might have occurred in the hours or days preceding the fire; severe arguments or threats of arson; and so on.

Most people are quite willing to talk about the physical aspects of a fire, but less willing to divulge personal information. Occupants may hesitate to mention their smoking or drinking habits—or involvement with previous fires—even if this personal information has no connection with the fire.

Witnesses. People who witnessed some aspect of the fire—especially its early stages, before the arrival of emergency personnel—should be sought out and questioned. These witnesses can describe events that are critical to the establishment of the fire's origin. Anything they might have noticed, from the location of the first flames and smoke to unusual sounds or activity, can be important.

Perhaps the most important of these witnesses is the person who turned in the alarm—in most cases the first person to see the fire. If this person was a passer-by who happened to notice the fire, he or she may not have remained on the scene. (However, in many cases, such people do remain, to see what happens and perhaps to brag a bit—and so are available for interview.) If an occupant turned in the alarm, it is unlikely that that occupant would have set the fire deliberately. At any rate, the occupant would be available for questioning, as would a neighbor who first saw signs of fire and sounded the alarm. (Here, again, if a neighbor sounded the alarm while people within the building were unaware of the fire, those people would not seem to be suspect in an arson investigation.)

News Media Personnel. News reporters and photographers usually cover the more spectacular fires; however, many local newspapers and television stations send someone to all or most fires in their area of coverage. Reporters, photographers, and video cameramen are very reliable sources of information. They are all trained observers, and experienced reporters are extremely adept at questioning witnesses. Often these personnel will find, question, and photograph or film witnesses well before fire investigators can. The reporters and their photos, films, or videotapes (of witnesses, bystanders, and the fireground) often are able to provide valuable information. This information can be easily available to investigators if a spirit of cooperation exists between the fire department and the local news media.

News media personnel have the right to respond to and report on fires, provided they do not interfere with fireground operations. They will, of course, exercise that right when they see fit to do so. If department officials and officers cooperate by allowing reporters reasonable access to the fireground, the news media will probably reciprocate by providing the department with the information they have gathered. In some departments, the dispatchers advise local news media of the initial information concerning every alarm that is received. In return, media personnel have been very cooperative in sharing their information (and in other ways, such as publicizing anti-arson campaigns and making people aware of the contributions of the department to the community).

Indirect Information

Indirect information—concerning the fire building, its owners and occupants, its contents, and (perhaps) possible arson suspects—may be relatively unimportant when a fire is obviously of natural or accidental origin. The indirect information that may be needed will often be provided by the people who contribute direct information—occupants and neighbors who may have witnessed the fire, for example. If a fire is difficult to categorize, or is of suspicious origin, indirect information increases in importance; each additional bit of information about anything or anyone connected with the fire can add to the investigator's understanding of the events leading to the fire.

The people who are sources of indirect information concerning a fire are those familiar with the fire building, its occupants, and their routines and habits.

Neighbors. Residents of neighboring houses or apartments may be able to provide a wide range of information. If they are fairly friendly with the occupants of the fire structure or apartment, they may be able to describe the occupants' furnishings and room layout, personal habits, work history, character, personal problems, and home life. (Nosy, but not particularly friendly neighbors may also be able to provide much of this information.) Even a neighbor who knows little of the occupants' personal life may know something about their daily routine, and whether or not the occupants kept to that routine on the day of the fire.

Friends. Friends of the occupants can usually provide information of a personal nature—if they care to do so. A good friend might, for example, decide not to mention an alcohol or drug problem. Friends may also be knowledgeable concerning the possessions of the occupants; they may be consulted if robbery is suspected but the occupants are unavailable for questioning.

Relatives. Relatives may know everything or nothing about the occupants' personal life—and may or may not be willing to reveal what they know. In questioning relatives of the occupants (and friends and neighbors as well), investigators must remember that they are seeking only information relevant to the fire, and not the "juicy details" of someone's personal life. Moreover, unless the person being questioned is in police custody and has been made aware of his legal rights, he is under no obligation to answer questions. Any information he does provide to the investigator is *volunteered.*

Mail Carrier. The mail carrier is often quite knowledgeable about the neighborhood in which he delivers mail, and the families and businesses located there. By spending much of the day in the area, he gets to know the people and their routines, and he is usually aware of events that are out of the ordinary. He may also know the whereabouts of an occupant who has been away for several days. Yet he is frequently overlooked as a reliable source of information (Figure 9-1).

Police. Police officers can often provide indirect information, as well as information directly concerning the fire. They are, for example, aware of the reputations and, to an extent, personal habits of many citizens. While police personnel are concerned mainly with criminal involvement, they also know when there has been no hint of such involvement—that is, when a citizen has a "clean" record.

The information gathered from police sources must be recognized as background information only. A person without a police record can become an arsonist; similarly, a person whose home or business becomes involved with fire is not an arsonist simply because he has a criminal record.

Tradespeople. Local merchants, bankers, insurance agents, and even the people who deliver newspapers, groceries, and so forth, may be able to provide information concerning the habits and financial status of those connected with the fire. Usually, however, these witnesses would be questioned during an arson investigation rather than a fireground investigation—unless they are available on the premises at the time of the fire.

Maintenance and Security Personnel. In a multiple-unit dwelling, maintenance and security personnel get to know a great deal about the structure and the residents. They would be aware of

- Flammable materials that might be stored on the premises

Figure 9-1. *Valuable evidence can be obtained from a mail carrier.*

- Recurring electrical problems, such as a fuse that keeps blowing
- The details of the furnishing of most apartments
- The daily routines of building occupants
- Animosities between residents and the building owners or managers
- Personal details about many of the residents.

In a commercial or industrial building, maintenance personnel (many of whom work at night) can provide information concerning work routines, unusual occurrences or difficulties with mechanical or electrical equipment, the storage of dangerous and/or flammable materials, and the usual (and perhaps unusual) placement of contents. They most likely are aware of the condition of fire detection and suppression systems and may be responsible for their proper operation.

Watchmen and security guards are sometimes hired only to give the appearance that a building is being protected. However, a watchman or guard who knows his job—a retired police officer, for example—is a reliable source of information. A watchman can probably pinpoint the time of origin of the fire fairly accurately—based on when he made his various rounds of the structure. Security guards should have a list of persons who entered the premises after business hours, and the times at which they entered and left. They should know which parts of the premises were locked and which were occupied after hours. Both watchmen and security guards should also be able to tell, with reasonable accuracy, whether there was a change in the position, amount, or type of contents since their last work shift.

Employees and Executives of a Firm. After a fire in an industrial or commercial establishment, the investigator should interview several employees who work in the area where the fire started. These employees should be asked about the processes and materials connected with their work and the hazards involved. Information concerning minor fires that might have occurred in recent months, what caused them, how they were extinguished, and precautions taken to prevent fires would also be of value.

If the fire occurred during working hours, the employee who discovered it should be questioned in greater detail. Members of the plant fire brigade (if there is one) should be asked about the condition and operation of fire detection and suppresion systems.

Supervisors and executives should be able to supply information concerning manufacturing processes, materials, and their fire hazards, as well as fire protection systems. In addition, they should be aware of problems such as pilferage or theft, which the thief might have tried to cover with arson; threats to the company by former employees or by activist special-interest groups; labor problems; and any of dozens of other situations that occasionally become motives for arson.

INTERVIEWS

An interview involves obtaining information from a person who is not suspected of a crime. The information may be a volunteered statement or volunteered answers to questions. Since the interviewee is not in police custody, the requirements of the Miranda decision do not apply; the interviewee need not be formally apprised of his rights. An interrogation is the questioning of a suspect who is in police custody and who therefore must be told his rights.

In most jurisdictions, fireground investigators may conduct interviews but may not interrogate suspects. In an interview, the interviewee may provide as much or as little information as he chooses. Most people are willing to relate what they know, but occasionally a person will decide not to divulge any information. In the great majority of cases, this interviewee does not want to become involved, is trying to protect someone, or feels threatened for one reason or another. In other words, most of the people interviewed by fireground investigators are innocent of any crime. (In fact, when arson is a possibility, the investigator should avoid asking questions of any person whom he suspects of the crime; information volunteered by a suspect should, however, be carefully noted. The best course of action is to declare the fireground a crime scene and request that police detain the suspect and the witnesses until the arrival of arson investigators.)

The people who are interviewed may, however, have important information to contribute: That information is most dramatic when it concerns the crime of arson. But it may have far-reaching consequences when it concerns, say, the extreme flammability of some modern fabric, or the danger inherent in a new appliance. Whether or not arson is suspected, interviews are a major part of fireground investigation; they should be conducted as carefully as the examination of the fire structure.

Begin Interviewing Early

People whose statements or responses may bear directly on the investigation should be interviewed before they leave the fireground. Once they have left, they may be difficult to locate. Moreover, some people will talk freely at the fire scene, but become reticent when the excitement is over. Others may tend to forget or exaggerate details, even a short time after the fire. The sooner the occupants and witnesses are interviewed, the better the chance of obtaining clear, accurate statements.

Conduct Interviews in Private

Interviews should be conducted in a reasonbly quiet place, preferably away from crowds and eavesdroppers. This privacy helps to put the interviewee at ease, perhaps increasing his willingness to respond. It also minimizes the chance of interruption or misunderstanding of questions and statements. The investigator should be especially careful to ensure that persons who are still to be interviewed cannot hear what is being said; otherwise, their responses may be influenced by what they hear.

Have Someone Witness the Interview

Although privacy is important, one additional person should be present, if possible, to witness the statement and responses obtained by the investigator. This person should not be connected with the fire or the fire structure in any way, but should be a completely impartial witness to the interviews. The witness should be instructed to listen only, and not to interrupt or present his personal views.

Record Interviewees' Statements

The information volunteered during interviews should be recorded. If the statements and responses are recorded on tape, each interviewee should be asked to begin by giving his name and address; he should then be asked to state that he is volunteering his information, and that he knows his statements are being recorded. These facts should also be written in the investigator's notebook, along with the name and affiliation (or address) of the person witnessing the interview.

If a tape recorder is not being used, the investigator should make written notes of interviewees' key statements. Each interviewee should be identified in the notes, as should the witness to the statements.

Listen Carefully

The interviewee should be asked to state what he saw and heard. The investigator should then allow the interviewee to make his statement in his own words, at his own pace. The interviewee should not be rushed or forced into particular areas of inquiry. The investigator should, above all, be a good listener—and an accurate listener. As long as the interviewee is speaking, the investigator should refrain from interrupting and should ensure that others do not interrupt. He should ask questions only when the interviewee seems hesitant.

Most of the time, the interviewee's statements and responses to questions will be direct and honest. However, a series of evasive answers, or statements that are in conflict with physical evidence such as burn patterns, may cause the investigator to become suspicious. The investigator should not, under any circumstances, accuse the interviewee (or anyone else) of arson; that might jeopardize both the information gathered during the interview and a possible case against the arsonist. Instead, if the investigator begins to suspect the person being interviewed, he should bring the interview to a close. He should then establish the fireground as a crime scene and inform the proper law enforcement agency of his suspicions. They will take the suspect into custody as prescribed by law and proceed with a criminal investigation.

If the interviewee confesses voluntarily, the investigator should allow him to complete his confession. Then the investigator should immediately inform law enforcement personnel that he has received a voluntary confession, that he is declaring the premises a crime scene, and that he will transfer responsibility for the investigation to their agency.

Ask Questions Only As Necessary

The proper and effective use of questions can help interviewees collect their thoughts, recall what took place, and give a complete account of what they saw and heard. However, questions can also interrupt the train of thought of a person who is making a statement. If an interviewee seems unable to begin, one or two questions may help him to start talking. And in some cases a person will experience difficulty in expressing himself and will need help from the investigator, in the form of questions. Otherwise, questioning should wait until the interviewee has described the incident in his own words.

When the investigator does begin to ask questions, he should use them to clarify the interviewee's statements or to elicit additional information. Questions should be addressed only to the objective of the interview—to determine when and how the fire started.

Interviewing and questioning techniques usually improve with experience. Investigators may also practice on each other, using the role-playing training method. Four basic rules regarding questions should be kept in mind:

1. *Ask simple questions.* Questions should be direct and to the point. An investigator who wants to know the color of the flames that were observed by the interviewee should ask, simply and directly, "What color were the flames?" There is no need to preface the question with a long explanation of the importance of flame color. It would probably confuse the interviewee or distract him.
2. *Ask one question at a time.* A long question that is really a string of shorter questions is confusing; some parts will probably be forgotten and never answered. If several questions need to be asked about a particular situation, the investigator should ask one question, and then wait for an answer; then ask the next question and again wait; and so on until he has all the information he requires.
3. *Phrase questions so that they do not influence the response.* The way a question is asked can lead the interviewee to a particular answer or to a choice between certain answers. For example, the question "Was the strange odor more like gasoline or kerosene" would probably be

answered with one of the two choices; yet the interviewee might have been thinking the odor was that of lighter fluid. A better question is, "What did it smell like?"

4. *Avoid questions with yes-or-no answers.* Questions of this type are sometimes useful, but they tend to limit the response. The investigator asks questions in order to draw out the interviewee; he will usually obtain much more information from a descriptive answer than from a yes or a no. For example, in answer to the question "Did you see flames at the second floor?" the interviewee would probably answer "yes" or "no" and then stop to wait for the next question. The question "Where did you see flames" might elicit an answer that contains entirely new and unexpected information.

Review Each Interview

At the end of each interview, but while the interviewee is still available, the investigator should take a few moments to review his notes. He should make sure he understands the interviewee's statement, and ask for clarification if necessary. He should also, at this time, try to resolve conflicts in the statements and responses of interviewees. Eventually, the investigator will have to sort out conflicting statements; it is easiest to do so by asking additional questions.

The investigator should realize, however, that conflicting statements are almost inevitable in the circumstances surrounding a fire. The excitement tends to lead to honest misconceptions or misinterpretations; what people think (and say) they saw is not always what they actually saw. This is one reason why testimonial evidence must be supported by physical or demonstrative evidence whenever possible.

Corroborate Important Points

When important information is provided by an interviewee, the investigator should attempt to corroborate it. Some of this information can be confirmed by physical evidence—the smell of gasoline, for example, can be confirmed by a residue found in the fire structure. Other information can be confirmed only by another person.

Corroboration is not achieved by remarking to an interviewee that "Smith heard two explosions; how many did you hear?" Instead, the question should be posed so as not to influence the response.

Checklist of Information

The list that follows includes most of the items of information that may be obtained through interviews. Some of the items may not apply to specific situations—but the general categories of information should be touched on during each interview.

Origin of the Fire. An interviewee who was present when the fire began may be able to discuss:

- Where the fire started
- How the fire started
- What material was the first to show flames
- The colors of the flames and smoke.

An interviewee who did not see the fire start should be asked:

- Where the first signs of fire were seen
- What they were.

Actions Taken. Interviewees should be able to discuss what they did after realizing there was a fire. Possible actions include:

- Turned in the alarm
- Left the building
- Helped others out of the building
- Attempted extinguishment
- Watched fireground activities.

Sounds. Among the sounds that interviewees may have heard at the time of the fire are:

- Breaking glass
- Explosions
- Shouting and/or arguments
- Vehicles leaving the area
- Dogs barking.

Odors. Interviewees who were in or near the fire building may have noticed the odors of:

- Gasoline, kerosene, or other flammable liquids
- Gaseous fuels
- Fireworks or cordite (acrid odors).

Unusual Actions. Interviewees who were in or near the fire building may have seen:

- Persons leaving the building just before the fire started
- Persons running from the scene
- Vehicles leaving the scene.

Unusual Conditions. Those in or near the fire building may also have noticed:

- Strangers in the vicinity of the building
- An unfamiliar vehicle parked near the building
- Activities not in keeping with the usual routine.

Special Contents. Occupants and others familiar with the premises should be asked about the presence of such contents as:

- Valuables, including paintings, coin and stamp collections, jewelry, oriental rugs, antiques
- Firearms, ammunition, powder
- Flammable liquids
- Chemicals.

Occupants. Primarily in the case of dwelling fires, the following information concerning the occupants could be related to possible reasons for the fire:

- Heavy smoking
- Heavy drinking
- Use of drugs
- Poor housekeeping habits
- Employment
- Stability (quick-tempered versus complacent).

The combination of heavy smoking and heavy drinking (or drug use) has led to many fires. Lack of employment—especially for some time—might motivate a person to set a fire to collect insurance money. And persons who are quick-tempered are more likely than even-tempered people to set fires in anger or as the result of an argument.

INFORMATION FROM RECORDS

The information contained in records is mainly indirect information that does not deal with the fire as such. However, like the indirect information obtained during interviews, it can be crucial to the investigation.

Fire Department Records

Fire department preplans and fire prevention and inspection records contain a good deal of information concerning industrial and commercial sites. This information should include:

- Building construction details, fire zones, locations of vertical and horizontal openings
- Details of fire detection and protection systems, including operating information and diagrams showing protected areas
- Descriptions of manufacturing processes and inherent fire hazards
- Normal positions of contents
- Locations where hazardous chemicals or flammable liquids are stored
- Reports of previous fires at the site, including the results of fireground investigations of those fires.

The first five items, in conjunction with the interior examination, can indicate whether there was anything unusual about the building or the behavior of the fire. The last item may point to hazardous conditions that were not rectified, or perhaps indicate that the building was the target of more than one arson fire.

Fire department records of previous fires involving the same fire building or the same owners or occupants can be enlightening, especially if they indicate similar circumstances or patterns. An individual who got away with arson once might very well be tempted to try it again.

Photographs and Videotapes

The photos, films, and videotapes produced by news media personnel, by spectators, and by fire department photographers (other than those working with investigators) are additional sources of information concerning both the fire and events occurring on the fireground. Photos of spectators, in particular, have been of help to investigators in many instances of arson. (Some departments provide a camera for the cab of every pumper. After those lines are supplied, the pump operator uses the camera to photograph the crowd—as unobtrusively as possible—and fireground operations.)

Photographs, films, and videotapes that are borrowed from news media or spectators (as well as those that are the property of the fire department) may be used as documentary evidence. For that purpose, they must be identified, marked, and maintained within the chain of custody as detailed in Chapter 8.

Other Records

Businesses. Insurance company fire inspection records can supply information concerning operating procedures, requests to rectify unsafe conditions, and recommendations for additional safety devices or systems.

The records of supervisory companies such as ADT and Wells Fargo can supply data concerning the maintenance and testing of fire protection systems, as well as information about recent attempts to disrupt such systems or to gain illegal entry to the fire building.

Records indicating heavy financial losses (sometimes a motive for arson) may be found in the fire structure. The civil courts may have records of suits brought against the business or its owners; creditors, including banks, may provide evidence that the business is deeply in debt.

Dwelling Units. When the occupant of a residence is unavailable and unknown, his identity may be found from:

- Electric and gas bills, which are almost always sent in the name of the occupant

- Telephone bills, which can provide the name of the occupant as well as long-distance numbers called by the occupant
- The occupant's telephone/address book, appointment calendar, etc., which may contain the names of close friends or relatives.

The occupant's wallet or purse, if available, may contain a driver's license, credit cards, membership cards, or other means of identification. Before a wallet or purse is opened, however, a police officer should be asked to observe what, if anything, is taken. The wallet or purse should then be transferred to the police, and a signed receipt obtained.

The records kept by motels, hotels, nursing homes, and hospitals may list such items of information as the names, permanent addresses, and next of kin of guests and patients. The records of many apartment houses contain similar information concerning residents, and may also include employment and credit data.

10

VEHICLE FIRES

Vehicle fires account for millions of dollars of damage each year. Because each individual loss is relatively small, vehicle fires are rarely publicized. For the same reason, vehicle fires are almost automatically reported as due to careless smoking or to an undetermined cause. Even when a vehicle fire is obviously arson, only minimal effort is expended to apprehend the arsonist.

This offhand attitude toward vehicle fires produces several unfortunate results:

- Young firesetters and disturbed arsonists often begin their arson "careers" with vehicle fires. When they are not apprehended, they may move up to structure fires, where the property damage is much greater and there is a possibility of causing injury or death.
- The incidence of vehicle arson fires increases as more and more one-time arsonists come to believe they can get away with setting fire to their vehicles for gain or perhaps to get rid of a gas guzzler. In some areas, professional vehicle arsonists have set up businesses; they will arrange, with the owner, to "steal" and torch his unwanted car or truck for a fee. In addition, vehicle arson is used as a weapon for revenge, as a fairly safe (although certainly illegal) tactic against business rivals, and often as the finale to a joyride in a stolen car.
- Mechanical and structural causes of accidental vehicle fires—often the types of vehicle fires that cause deaths—remain undiscovered, because the involved vehicles were not carefully examined.
- The increasing incidence of vehicle arson results in higher insurance rates for all vehicles.

Vehicle fires can be investigated as effectively as structure fires. The principles are the same for both types. Moreover, most vehicle fire investigations can be completed in a reasonably short time, so they do not add greatly to the work load of investigators or investigation teams. And, as is true for structure fires, fireground investigation is the key to reducing the incidence of arson fires involving vehicles.

INVESTIGATION OF VEHICLE FIRES

The objective in investigating a vehicle fire is, as usual, to determine the point of origin, heat source, reason, and category of the fire. The procedure, once again, is to perform first an exterior examination and then an interior examination. The exterior examination includes both the area surrounding the involved vehicle and the exterior of the vehicle itself. The interior examination includes the engine compartment, trunk, truck bed or body, and driver/passenger compartment. In addition, the observations of firefighters and statements of witnesses are solicited.

The discussions of notes, photos, diagrams, and the collection and preservation of evidence (Chapters 6 and 8) apply to vehicle fires as well as structure fires. The front, back, and sides of the vehicle should be photographed, as well as the interior burn patterns; close-up photos of the license plates and vehicle identification number should also be taken. The location of the involved vehicle may be established by including nearby street signs, road markers, or other stationary features in one or two photos.

Special Considerations Regarding Vehicle Fires

Although the principles of fireground investigation apply to both structure and vehicle fires, there are differences between the two. These differences stem mainly from the mobility of vehicles and the presence of a flammable liquid in almost every vehicle. They will sometimes indicate how extensive an investigation should be.

Collision. When a vehicle fire is ignited by a collision, there is little chance that arson is involved. The reason and category of the fire are obvious, and a detailed investigation is unnecessary (except, perhaps, to gather information concerning vehicle safety). The involved vehicle or vehicles should, however, be examined to confirm that it was the collision that caused the fire.

Gasoline and Other Flammable Materials. Most cars and trucks run on gasoline and carry a tank of this flammable liquid. (Diesel fuel is much less flammable than gasoline.) In addition, some drivers carry a container of spare gasoline in their vehicles. Flammable liquids of many varieties are transported in trucks, and tanks of flammable cooking gas are carried on many recreational vehicles. All this adds up to quite a bit of flammable material moving along the highways. A sudden stop or a crash could cause a container of flammable liquid to overturn, ignite, and perhaps involve the entire vehicle, including the gas tank. Or a collision would rupture the gas tank, causing a fire that soon involves any flammable liquids that are being transported. The investigator must be careful to accurately establish the cause of the fire in such cases.

There have been a number of court cases involving rear-end collisions in which a vehicle's gas tank was ruptured, causing fire and injury or death. If a container of gasoline was stored in the trunk of a car involved in a rear-end collision and ensuing fire, then

1. The container may have ruptured, causing a fire that spread to the gas tank and its contents.
2. The gas tank may have ruptured, causing a fire that spread to the container.
3. Both the container and the gas tank may have been ruptured in the collision, with both contributing to the fire.

The investigator's determination of the point of origin, heat source, and reason for the fire then become vital to the disposition of legal claims.

There is another common cause of vehicle fires that involves gasoline. In cold climates, drivers sometimes pour gasoline into the carburetor throat to get the engine started. Occasionally the gasoline backfires, causing a flash of flame to shoot up from the carburetor. The flame may startle the driver, who will drop the gasoline container or spill gasoline on the engine. The result is an extensive vehicle fire.

Location. The investigator should be wary of arson when a burned or burning vehicle is discovered in a remote area—especially when the driver is not present. When the driver turns in the alarm at the outset of the fire, however, it is probable that the fire is accidental.

Truck cargo fires are occasionally discovered on the highway, while the truck is in motion. Such a fire may or may not be deliberately set, but the driver cannot logically be excluded as an arson suspect.

Careless Smoking. As with structure fires, careless smoking is a common reason for vehicle fires. However, in most cases there is a long period of smoldering preburn before open flames are produced. Smoke production during the preburn gives ample warning to the vehicle's occupants, and the fire is usually discovered and extinguished before flames appear. If the smoldering fire is not detected and the vehicle is parked where it is obscured from view, the fire can become quite extensive before it is discovered. Such fires have, for example, involved other vehicles in a garage—and the garage itself.

Recreational Vehicles. Live-in type travel trailers are, essentially, small dwellings on wheels that are pulled by a powered vehicle. A fire in this type of unit should be investigated as a dwelling fire. (This is also true for mobile homes—which are not really mobile—as implied in the chapters on structure fires.) Fire in the powered vehicle is, of course, investigated as a vehicle fire.

Motor homes are small dwellings on truck chassis. A fire within the live-in section should be investigated as a dwelling fire, but complications due to the engine and fuel supply (including cooking fuel) should be considered; a fire in the truck or powered part of a motor home should be investigated as a vehicle fire. In most cases, both will have become involved—especially when the fire occurs on a limited-access highway or a remote campsite, so that a fairly long time is required for fire companies to arrive.

Trucks. A fire involving the cab, engine compartment, or underbody of a truck is investigated as a vehicle fire. Enclosed cargo compartments (especially the trailer of a tractor-trailer combination) are like small warehouses and should be investigated as such.

The possible heat sources in large trucks include spontaneous ignition of cargo, the interaction of noncompatible materials, faulty electrical wiring and components, malfunctioning refrigeration units, and friction. As with other types of vehicles, careless smoking is responsible for many truck fires. A collision may cause an extensive fire, especially when flammable cargo is being transported.

Large trucks may be the targets of arson fires for all the reasons that pertain to business structures. When a truck burns, both the truck and its cargo become the objects of insurance claims. Obsolete inventory may be torched in a truck, to ensure that the plant or warehouse and its salable contents are not harmed. Or a trucker may steal part or all of a valuable load, perhaps replace it with goods of little value, and then set fire to the truck to hide the theft. And hijacked trucks are sometimes set on fire after the load is removed. Police and fire marshals can usually discover such motives for arson. The fireground investigator's task is to determine the category of the vehicle fire, and to alert the proper authorities when arson is suspected.

CASE HISTORY: Semi-Trailer Fire

Fire companies arriving at the scene of a semi-trailer fire on an interstate highway found the trailer fully involved. Firefighters had some difficulty in keeping the fire extinguished, and they observed other abnormal conditions:

Observation	*Evaluation*
• Fully involved trailer	• Rapid fire spread and/or delay in reporting the fire

• Orange-yellow flames mixed with blue flames	• Burning of ordinary combustibles along with gases of some kind
• Difficulty in extinguishing fire	• Typical of flammable liquid involvement
• Occasional small explosions followed by increased burning intensity	• Containers of flammable liquid?
• Unusual odor	• Flammable liquid again?

The fireground investigator arrived at about the time the fire was finally extinguished. He found no point of origin in the trailer, but rather a large area of burning. The trailer load was comprised of cases of alcoholic beverages, which accounted for the blue flames, small explosions, and unusual odor. The driver stated that he had been driving along the highway and noticed the fire in his rear-view mirror. When he pulled over and opened the trailer doors, he found the whole interior on fire.

The investigator made a quick approximate check of the remains of the trailer contents against the bill of lading. He then put in a call for a fire marshal, who impounded the trailer and had it hauled off the highway. With a representative of the company that owned the cargo, the fire marshal then checked the contents, bottle by bottle, against the bill of lading. Although many bottles were broken and their plastic caps were melted, the caps could be counted. It was found that about half the load was missing.

The driver was the obvious suspect. Under interrogation he admitted to stopping along his route, unloading half the cargo with two partners, and then driving back onto the interstate. There, he set fire to the remainder of the load, using crumpled paper and liquor from some of the bottles.

EXTERIOR EXAMINATION

Before beginning the exterior examination, the investigator should survey the scene, as he would in the case of a structure fire. He should observe whether the ground area around the fire vehicle is soft enough to hold footprints or tire tracks (in which case he may want to rope off the area to preserve this evidence); whether grass, brush, or trees seem to have been involved; the general position of the vehicle relative to roadways; and the presence of vehicle parts or containers near the fire vehicle. If the fire involved a truck that is known to have been hijacked, the fireground is a crime scene; it should be cordoned off immediately to prevent the destruction of evidence, especially the tire tracks of a vehicle that may have been used to unload the truck.

Examining the Ground Around the Vehicle

The investigator should begin with a slow tour around the fire vehicle, recording any information that appears significant. Vehicle parts found nearby might indicate that the fire vehicle was first "stripped" and then burned. Containers in the area might have been used to carry an accelerant. These items should be preserved in place for later collection as evidence.

Soft Ground. If the fire vehicle is situated on a hard surface, such as concrete or macadam, there may be little else to observe. On soft ground, there may be tire tracks or footprints, other than those of firefighters. Footprints should be preserved, to be compared with those of suspects if the fire is determined to be an arson fire. Tire tracks, if they do not belong to the fire vehicle, might indicate that a second vehicle was used to tow or push the fire vehicle to the site. This might be the case if the engine of the fire vehicle needs major repairs; the owner might have decided to both avoid the repairs and collect the insurance money by setting fire to the vehicle. Or the second set of vehicle tracks might belong to a car or truck that was used to flee the scene. In any case, tire tracks should be preserved if they do not seem to match those of the fire vehicle.

Involved Grounds. When grass, brush, or trees in the surrounding area are burned, the fire may have spread from them to the vehicle, or vice versa. The wind direction at the time of the fire and the burn patterns are of help in determining which actually occurred, as is the case of a structure fire (Chapter 6).

When fire extends (with the wind) along grass or brush to a vehicle, it leaves a burn pattern along the ground. In addition, the burn pattern on the vehicle exterior indicates upward flame travel, beginning at the point where the flames moved from the ground to the vehicle. Fire extending from an involved vehicle to grass will fan out and leave a wide burn pattern on the grass.

A "clever" arsonist might start a fire in grass or brush, and then position the vehicle in its path. There are two ways to see through this ploy. First, the point of origin of the original grass or brush fire must be located and examined. The investigator should be able to find a heat source—a barbecue grill, a campfire, or even a chunk of broken glass that could have focused the sun's rays on the grass—at the point of origin. If no obvious heat source can be found, then arson should be suspected. Second, an arsonist would probably use an accelerant to make sure the fire moves quickly and in the proper direction. Residues of flammable liquids may be found on the ground near the point of origin. In addition, the investigator should check the path of the ground fire; flammable liquids cause distinctive burn patterns in grass, just as they do in rugs and carpets. Again, anything out of the ordinary should suggest arson.

The ground area under the vehicle may be examined at this time or when the exterior of the vehicle is examined.

Examining the Exterior of the Vehicle

Probably the first item that should be checked is the vehicle identification. The license plates are the most obvious identification, but in cases of arson they may be missing or they may have been changed by a thief. If the license plates are available, their number should be recorded.

Every vehicle sold in the United States has its own vehicle identification number (VIN). It is recorded when the vehicle is registered or sold, and it appears on the vehicle registration or title certificate. It also is marked on the vehicle itself. On American-made cars, the VIN is on a plate located on the driver's side of the dashboard, where the dashboard and windshield are joined. The VIN can be read from outside the vehicle and can be used to identify the vehicle and its owner. It, too, should be recorded by the investigator. Other VIN locations are shown in Figure 10-1a and b.

The next step is to examine the outside of the vehicle carefully, observing first the burn patterns and then specific items or parts of the vehicle.

Exterior Burn Patterns. During a second tour of the vehicle, the investigator should examine the burn patterns on the vehicle's exterior paint (Figure 10-2). The exterior burn patterns should be compatible with the wind direction and force at the time of the fire. If a flammable liquid was spilled or splashed on the outside of the vehicle, the burn patterns may be erratic or may indicate fire spread *into* the wind. In addition, the low burn and the point of origin may be indicated by blistered and peeled paint, even if the fire was confined to the interior of the vehicle.

Roof. At the same time, the roof should be checked for sagging (Figure 10-3). The most pronounced sagging will be located directly over the hottest part of the fire—usually the point of origin. The sagging occurs because the metal roof and its supports lose strength at elevated temperatures. The higher the temperature, the greater the loss of strength and the more pronounced the sagging.

Gas Tank Filler Tube. The metal fuel tanks used in cars and trucks are quite sturdy. They rarely develop leaks unless they are damaged; and, when a tank does develop a leak, it is repaired immediately to eliminate the danger of explosion and the waste of gasoline. Unless a tank is ruptured in a collision, it will rarely explode or add fuel to a fire involving the vehicle.

The supply of flammable liquid contained in gas tanks is, nonetheless, very tempting to arsonists, who may use any of several methods to involve

FOREIGN PASSENGER VEHICLES

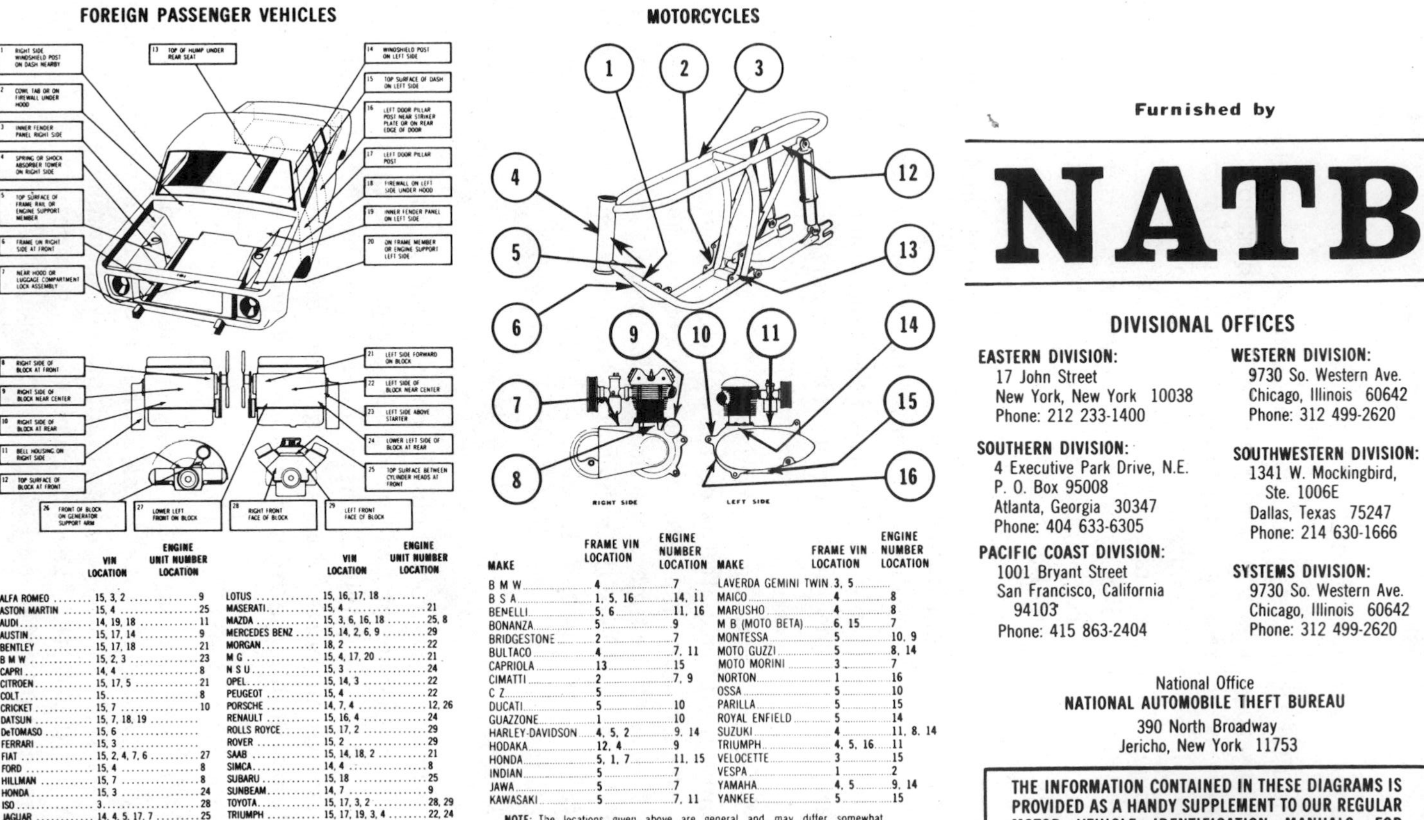

	VIN LOCATION	ENGINE UNIT NUMBER LOCATION
ALFA ROMEO	15, 3, 2	9
ASTON MARTIN	15, 4	25
AUDI	14, 19, 18	11
AUSTIN	15, 17, 14	9
BENTLEY	15, 17, 18	21
B M W	15, 2, 3	23
CAPRI	14, 4	8
CITROEN	15, 17, 5	21
COLT	15	8
CRICKET	15, 7	10
DATSUN	15, 7, 18, 19	
DeTOMASO	15, 6	
FERRARI	15, 3	
FIAT	15, 2, 4, 7, 6	27
FORD	15, 4	8
HILLMAN	15, 7	8
HONDA	15, 3	24
ISO	3	28
JAGUAR	14, 4, 5, 17, 7	25
JENSEN	15, 17	8
LAMBORGHINI	15, 17, 19	25
LANCIA	3, 4	9
LOTUS	15, 16, 17, 18	
MASERATI	15, 4	21
MAZDA	15, 3, 6, 16, 18	25, 8
MERCEDES BENZ	15, 14, 2, 6, 9	29
MORGAN	18, 2	22
M G	15, 4, 17, 20	21
N S U	15, 3	24
OPEL	15, 14, 3	22
PEUGEOT	15, 4	22
PORSCHE	14, 7, 4	12, 26
RENAULT	15, 16, 4	24
ROLLS ROYCE	15, 17, 2	29
ROVER	15, 2	29
SAAB	15, 14, 18, 2	21
SIMCA	14, 4	8
SUBARU	15, 18	25
SUNBEAM	14, 7	9
TOYOTA	15, 17, 3, 2	28, 29
TRIUMPH	15, 17, 19, 3, 4	22, 24
T V R	15	
VAUXHALL	15	22
VOLKSWAGEN	15, 13, 7	12, 26
VOLVO	14, 2	22

NOTE: The locations given above cover a wide span of year models. Little consistency was found in the location of the VIN plate in the older models.

MOTORCYCLES

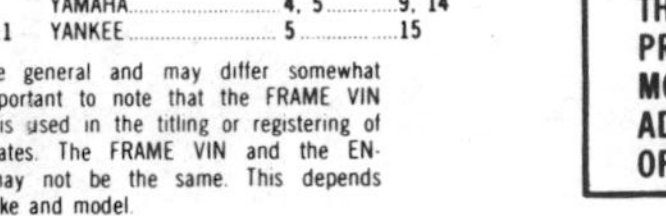
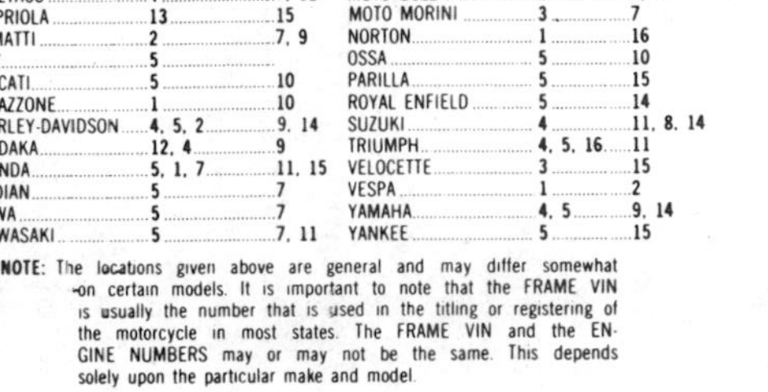

MAKE	FRAME VIN LOCATION	ENGINE NUMBER LOCATION
B M W	4	7
B S A	1, 5, 16	14, 11
BENELLI	5, 6	11, 16
BONANZA	5	9
BRIDGESTONE	2	7
BULTACO	4	7, 11
CAPRIOLA	13	15
CIMATTI	2	7, 9
C Z	5	
DUCATI	5	10
GUAZZONE	1	10
HARLEY-DAVIDSON	4, 5, 2	9, 14
HODAKA	12, 4	9
HONDA	5, 1, 7	11, 15
INDIAN	5	7
JAWA	5	7
KAWASAKI	5	7, 11
LAVERDA GEMINI TWIN	3, 5	
MAICO	4	8
MARUSHO	4	8
M B (MOTO BETA)	6, 15	7
MONTESSA	5	10, 9
MOTO GUZZI	5	8, 14
MOTO MORINI	3	7
NORTON	1	16
OSSA	5	10
PARILLA	5	15
ROYAL ENFIELD	5	14
SUZUKI	4	11, 8, 14
TRIUMPH	4, 5, 16	11
VELOCETTE	3	15
VESPA	1	2
YAMAHA	4, 5	9, 14
YANKEE	5	15

NOTE: The locations given above are general and may differ somewhat on certain models. It is important to note that the FRAME VIN is usually the number that is used in the titling or registering of the motorcycle in most states. The FRAME VIN and the ENGINE NUMBERS may or may not be the same. This depends solely upon the particular make and model.

Furnished by

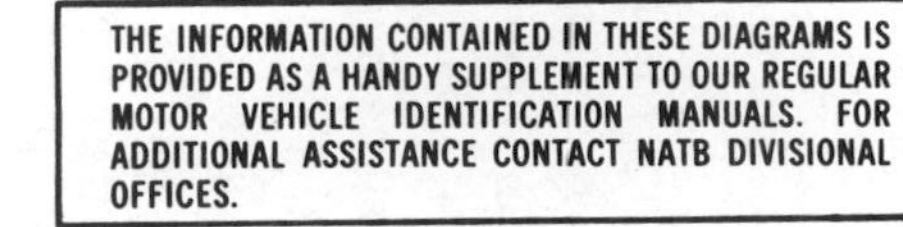

DIVISIONAL OFFICES

EASTERN DIVISION:
17 John Street
New York, New York 10038
Phone: 212 233-1400

SOUTHERN DIVISION:
4 Executive Park Drive, N.E.
P. O. Box 95008
Atlanta, Georgia 30347
Phone: 404 633-6305

PACIFIC COAST DIVISION:
1001 Bryant Street
San Francisco, California
94103
Phone: 415 863-2404

WESTERN DIVISION:
9730 So. Western Ave.
Chicago, Illinois 60642
Phone: 312 499-2620

SOUTHWESTERN DIVISION:
1341 W. Mockingbird,
Ste. 1006E
Dallas, Texas 75247
Phone: 214 630-1666

SYSTEMS DIVISION:
9730 So. Western Ave.
Chicago, Illinois 60642
Phone: 312 499-2620

National Office
NATIONAL AUTOMOBILE THEFT BUREAU
390 North Broadway
Jericho, New York 11753

THE INFORMATION CONTAINED IN THESE DIAGRAMS IS PROVIDED AS A HANDY SUPPLEMENT TO OUR REGULAR MOTOR VEHICLE IDENTIFICATION MANUALS. FOR ADDITIONAL ASSISTANCE CONTACT NATB DIVISIONAL OFFICES.

Figure 10-1a. *Front: Location of VIN numbers on various types of vehicles.*

AMERICAN PASSENGER VEHICLES

Starting [...] 1968, all domestic passenger cars were assembled with the VIN plate attached to the left side of the dash or instrument panel, visible through the windshield. There are two exceptions, one being the Corvette which carries the VIN attached to the left side windshield post. The other is the 1958 model Ford products which used the right side of the instrument panel.

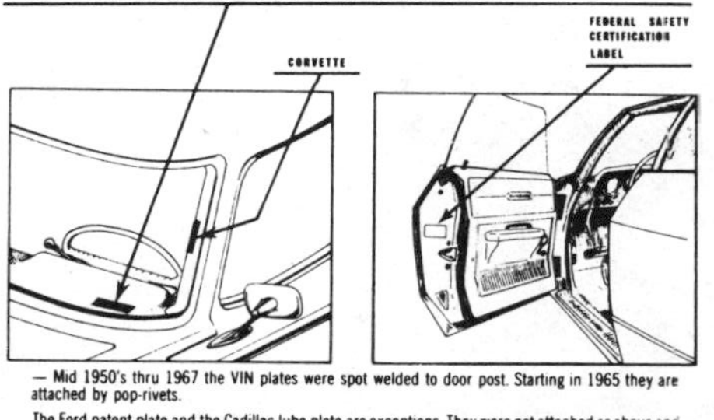

— Mid 1950's thru 1967 the VIN plates were spot welded to door post. Starting in 1965 they are attached by pop-rivets.

The Ford patent plate and the Cadillac lube plate are exceptions. They were not attached as above and should not be used. Only the die-stamped frame numbers under the hood should be relied upon.

Vehicles prior to 1956 were identified by the motor number.

ENGINE AND TRANSMISSION NUMBER LOCATIONS

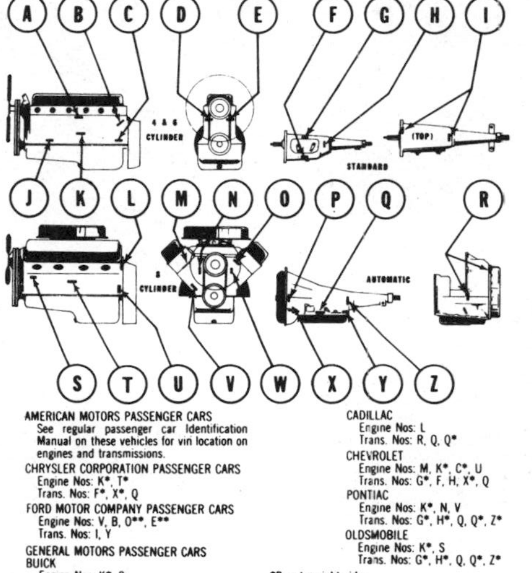

AMERICAN MOTORS PASSENGER CARS
See regular passenger car Identification Manual on these vehicles for vin location on engines and transmissions.
CHRYSLER CORPORATION PASSENGER CARS
Engine Nos: K*, T*
Trans. Nos: F*, X*, Q
FORD MOTOR COMPANY PASSENGER CARS
Engine Nos: V, B, O**, E**
Trans. Nos: I, Y
GENERAL MOTORS PASSENGER CARS
BUICK
Engine Nos: K*, S
Trans. Nos: G*, H*, Q, Q*
CADILLAC
Engine Nos: L
Trans. Nos: R, Q, Q*
CHEVROLET
Engine Nos: M, K*, C*, U
Trans. Nos: G*, F, H, X*, Q
PONTIAC
Engine Nos: K*, N, V
Trans. Nos: G*, H*, Q, Q*, Z*
OLDSMOBILE
Engine Nos: K*, S
Trans. Nos: G*, H*, Q, Q*, Z*

*Denotes right side
**Denotes rear face of block

COMMERCIAL MOTOR VEHICLES

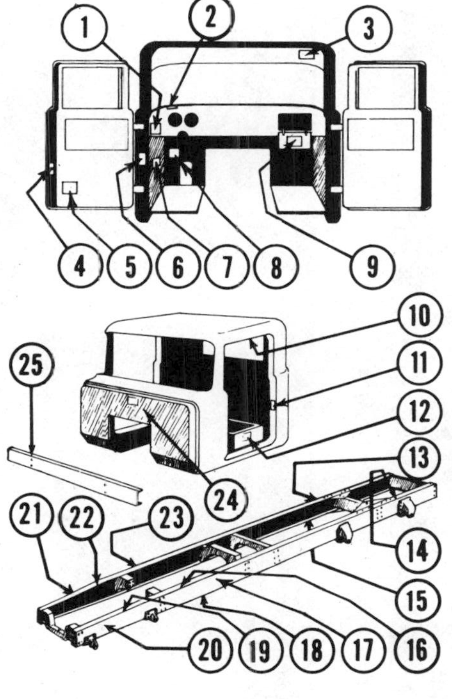

MAKE	VIN PLATE Location	POLICE VIN Location
AUTOCAR	5	20
BROCKWAY	5	23
C C C (Crain Carrier)	5	
CHEVROLET		
Truck Tractors	11	16, 18
Light Duty	11, 10	19
LUV Pickup	2	
El Camino	2	
DATSUN	11	24
DIAMOND REO	4, 5, 12	17
DIVCO	3	
DODGE		
Truck Tractors	11	21
Light Duty	11	21
FORD		
Truck Tractors	4, 9	13, 15
Light Duty	4, 9	22
COURIER Pickup	2	
RANCHERO	2	
G M C		
Truck Tractors	11, 12	16, 18
Light Duty	11	22
SPRINT	2	
INTERNATIONAL		
Truck Tractors	4, 5	17, 20
Light Duty	4, 5	20
JEEP	6, 24	19
KENWORTH	5	25
MACK	5	23
MARMON	1	14
MAZDA	11	24
OSHKOSH	5	21
PETERBILT	7, 8	17, 23
PLYMOUTH	11	
WHITE TRUCKS	5	21, 24
WHITE FREIGHTLINER	5	20, 22

GENERAL TRUCK-TRACTOR IDENTIFICATION INFORMATION

• The manufacturers of truck-tractors have not, to date, developed a totally uniform identification system.

• There is no uniformity as to the utilization of Alpha and Numeric characters in the configuration of vehicle identification numbers for the various makes of truck-tractors. These inconsistencies create a great demand for knowledge, diligence and thoroughness on the part of Motor Vehicle and Police personnel to insure that a full and correct vehicle identification number is always obtained and utilized for official purposes.

• As a general practice most all makes do have a manufacturers "serial plate" which usually provides (in addition to the vehicle identification number) considerable data as to the year, make, model and construction features. These plates, while not uniformly located, are most always attached to the cab (door, dash or seat base area). Attachment is usually by screws, brads or rivets. A majority of the manufacturers die stamp a repeat vehicle identification number into the frame rail.

Figure 10-1b. *Back: Location of VIN numbers on various types of vehicles.*

Figure 10-2. Examine burn patterns on exterior paint.

Figure 10-3. Sagging of the roof usually occurs directly above the hottest part of the fire.

the gasoline. Most of these methods don't work for the arsonist, but they do alert the investigator to the possibility of arson. The methods include:

- Removing the gas tank filler tube cap (gas cap)
- Stuffing rags in the filler tube
- Siphoning gasoline out of the gas tank for use as an accelerant
- Removing the gas tank drain plug
- Puncturing the gas tank.

The investigator should check the filler tube to see whether the gas cap is in place. If it is not, it may have been removed in the hope that an explosion would occur. (Actually, at most, some gasoline vapors might burn off at the filler tube opening; burning cannot occur within the filler tube, because the proper mixture of air and gasoline vapor cannot be achieved there.) The investigator should search the area for the missing gas cap. If it cannot be found, the investigator should try to determine why, by questioning the vehicle owner. If the gas cap is found undamaged at the scene, then it was removed from the vehicle before the fire. If it is found with distorted locking flanges, and the filler tube has matching damage (Figure 10-4), then the gas tank probably exploded and blew the gas cap off the filler tube. This can, of course, be checked by examining the gas tank.

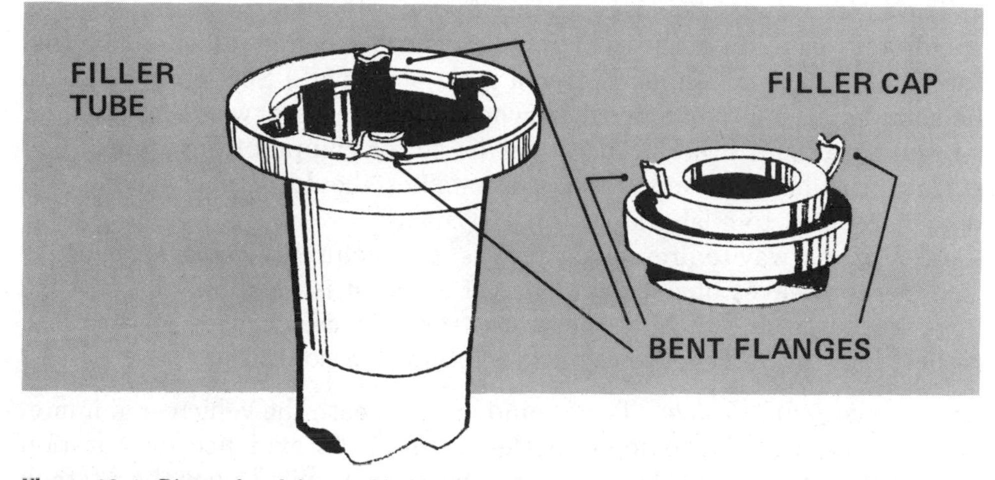

Figure 10-4. Distorted and damaged locking flanges of gasoline filler tube indicate explosion.

An amateur arsonist may stuff rags in the gas tank filler tube, in the hope that an explosion will occur. (Again, at most, there may be some burning at the filler tube opening—but no explosion. The material inside the tube will not even ignite.) However, when a gas cap is lost, the vehicle owner sometimes stuffs the filler tube opening with rags to keep the fuel from spilling or evaporating. When the investigator finds rags or other materials in the filler tube, he should, through questioning, try to determine how they got there.

Gasoline may be siphoned up out of the tank, through the filler tube, for use in accelerating a vehicle fire. Usually, this is difficult to detect—especially if the gas cap is replaced after the gasoline is siphoned out. However, a careless arsonist may leave the siphoning hose at the scene. Or, during the siphoning, he may splash gasoline on the side of the vehicle (this would be indicated by an erratic burn pattern), or spill some on the ground (which would be indicated by its residue).

Gasoline Tank. If possible, the amount of liquid in the vehicle's gas tank should be measured. If the tank is completely empty, the vehicle may have been towed to the scene; it may have run out of fuel, and then been stripped and/or burned while it was left unattended; or the gasoline may have been drained from the tank for use in accelerating the fire. The tank should be checked for signs that the fuel was removed.

A missing gas tank drain plug is an obvious indication that the tank was drained. The investigator should search for the drain plug washer and gasket, which are often overlooked when the drain plug is removed. If the plug is in place, but loose enough to be turned with the fingers, it was probably loosened to drain the tank and then replaced only finger tight. If the plug is screwed in tightly but shows fresh tool marks, an arsonist may have loosened it to drain the tank and then replaced it tightly; or, he may have tried to loosen the plug but found it frozen in place.

The investigator who suspects that fuel was drained from the gas tank should take samples of the road or ground surface beneath the tank. (Some fuel is almost always spilled when a tank is drained.) Soft ground should be sampled to depth of at least 2 inches (5 centimeters) with a heavy-duty garden type hand trowel. Pieces may be cut or chipped from a hard road surface with hammer and chisel. The samples should be packaged as evidence and sent to a laboratory for analysis.

An efficient way to drain a gas tank is to puncture it. However, because this method leaves obvious evidence of arson, it is used only by persons who are not trying to hide the crime—by thieves, perhaps, or vandals, rather than the vehicle owner or a professional arsonist.

Area Beneath the Vehicle. The ground area beneath the vehicle and immediately adjacent to it should be checked carefully for evidence of saturation with a flammable liquid. The liquid may have been spilled when the gas tank was drained (as noted above); it may also have dripped down through openings when it was thrown into the vehicle. If such signs as erratically burned grass or flammable-liquid residue are found, a sample of the ground or road should be collected for analysis. The laboratory may be able to determine whether the ground sample contains the same type and brand of gasoline.

Tires. When a vehicle is burned for insurance money, good tires are often removed and replaced with worn tires. The investigator should check to see whether this is the case. (Even if the tires are consumed or badly burned, the small section of each tire that is in contact with the ground will remain intact. If necessary, the vehicle should be rolled forward a bit so that these areas may be inspected.) (Figure 10-5 and Figure 10-6.)

If the tires are obviously not roadworthy, a sample of one tire should be collected as evidence. If the tires are missing, the fire is obviously of suspicious origin.

Hubcaps and Wheels. As with tires, cheap hubcaps and wheels may be substituted for expensive wheel covers or "mag" wheels before a car is burned. Usually, this is the work of the owner rather than thieves, so ques-

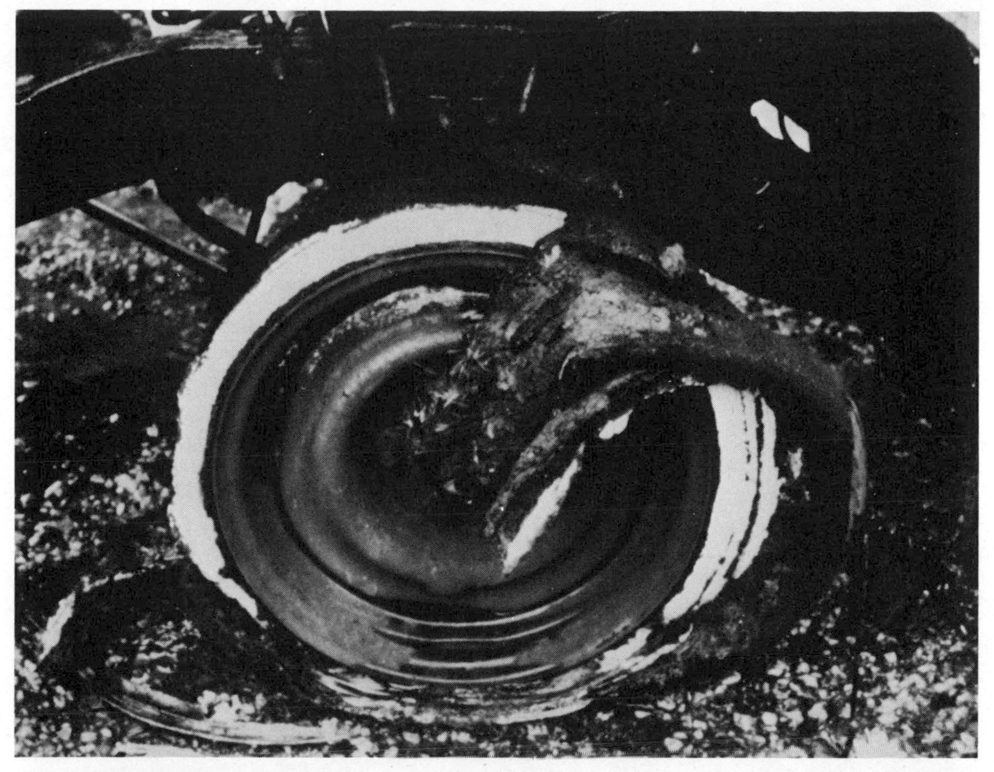

Figure 10-5. *Examine tire surface.*

Figure 10-6. *This tread was protected from flame contact and can be measured.*

tioning the owner about replaced tires, wheels, and hubcaps will do little good. To check for recent replacement of these items, the investigator should examine the lug bolts. If the wheels have been removed and replaced, the dirt and grime around the lug bolts will be disturbed; there will be "clean" spots here and there. In some cases the arsonist, in his haste, may not have replaced all the nuts, or may not have tightened them all. Any of these indications may be cause for suspicion. Tool marks and "missing" road grime may show that tires or hubcaps have been replaced.

Doors. The investigator should question first-arriving firefighters concerning the positions of the vehicle's doors (and windows) on arrival. He should, in addition, inspect each door and its window.

Passengers fleeing a vehicle fire would most likely leave the doors open. An arsonist might open the vehicle doors to ensure a good draft. If first-arriving firefighters found more open doors than there were passengers, the investigator should try to determine why. For example, if there were only two passengers in a four-door sedan, those passengers would have had to open only two doors to escape the fire. If firefighters found all four doors open, the passengers must have opened the remaining two doors for some reason. That reason might be to allow oxygen in to the fire; however, it might also be simply to remove valuable items from the back seat before they were consumed.

Exterior burn patterns can also indicate whether the doors and windows were open or closed during a fire in the interior of the vehicle:

- If a door and its window were closed and the windows remained intact, there would be no exterior burn patterns on or above the door.
- A closed door with an open window would show burn patterns on the door above the window opening, and on the body above the door.
- An open door on the leeward side would allow the fire to burn the exterior paint above the doorway; the door itself would probably show no exterior damage. An open door on the windward side might exhibit no damage to the door or the doorway.
- If a pair of doors opposite each other were open, the point above the doorway on the leeward side would be heavily damaged; the doorway on the windward side would probably not be damaged at all.
- If the doors were closed but a pair of opposite windows were open, the door and doorway above the leeward open window would show heavy damage.

Additional signs indicating whether windows were open during the fire are discussed in the section on interior examination.

When the exterior of the vehicle is also involved, these patterns may be obscured by (or merged with) patterns due to the external burning.

Bumpers. To ensure that the vehicle is totally destroyed by fire, the arsonist may take it to a remote area, where it can burn undetected for a fairly long time. If the vehicle cannot be driven to the fire site, it may be towed or pushed there. Where circumstances suggest that this might have been done, the investigator should check both bumpers for fresh scratches or other indications of recent towing or pushing.

INTERIOR EXAMINATION

Each of the three interior spaces of the fire vehicle—the driver/passenger compartment, the engine compartment, and the trunk of an automobile—should be examined in turn. Although these spaces are quite small and confined, they can be investigated effectively, and burn patterns can be traced to the low burn and point of origin. The large enclosed cargo spaces on larger trucks are more like small rooms than vehicle spaces; they are examined in the same way as a room in a structure, as detailed in Chapters 7 and 8.

The discovery of a low burn and point of origin in one of the three interior spaces should not keep the investigator from continuing the interior examination. Although these spaces are separated from each other, fire can extend from one to the other. Moreover, what is found in one space may indicate an accidental fire, while the evidence in another compartment may point to arson. The determination of the category of a vehicle fire most often depends on the combined results of the exterior examination, the examination of all interior spaces, and the statements of the vehicle owner and witnesses to the fire.

Driver/Passenger Compartment

The driver/passenger compartment should first be examined to establish the low burn and locate the point of origin—if there is one in that compartment. Then the space should be checked for signs indicating how the fire started and spread.

Burn Pattern. The areas nearest the low burn and point of origin are the most heavily damaged, in a vehicle as well as a structure. For example, a fire that originates on the driver's side of the vehicle will damage that side more heavily than the passenger side—whether the overall damage is relatively light or quite extensive. The burn pattern can usually be traced to the area of heaviest burning.

Vehicle door handles and window levers are made of white metal and plastic, with low melting points. The handles and levers nearer the low burn may be distorted, melted, or even totally consumed (Figure 10-7). Those further from the low burn may show only minor damage, or no damage at all. Upholstery, floor mats and rugs, and door liners nearer the seat of the fire will show heavier damage than those further from the point of origin. If the seat covering was consumed, the seat springs should be checked for elasticity, as described in Chapter 7; a lack of springiness and collapsed springs indicate proximity to the seat of the fire.

Figure 10-7. *Note the melted door handle nearest to the low burn.*

In addition, the sagging of the vehicle's roof, as observed during the exterior examination, will provide a clue as to the location of the point of origin.

It may be necessary to remove accumulated debris to find the true low burn, especially if the fire originated on the floor of the driver/passenger compartment. If the floor mats or rugs were destroyed, and the metal flooring is distorted or shows other signs of attack by heat, the point of origin may be beneath the vehicle.

In searching through or removing debris, the investigator should watch for fragments of window glass. These might indicate that a window was broken in by a thief (or arsonist) to gain entry (Figure 10-8). The remains of the vehicle's contents—metal buttons and zippers, for example, in the case of clothing—should also be observed and noted. Their presence would support the vehicle owner's claim that such items were destroyed by the fire. Their absence would point to either theft or false statements by the owner.

Windows. Windows should be checked from inside as well as outside. If the driver/passenger compartment was involved, unstained windows were probably open at the time of the fire. Heavy, thick interior staining indicates that the windows were closed and the heat built up slowly within the vehicle (Figure 10-9). Broken windows with fragments that are unstained and/or melted around the edges indicate the rapid buildup of intense heat—often a signal that accelerants were involved.

Figure 10-8. *Broken glass inside the vehicle may be reason to suspect unlawful entry.*

Figure 10-9. *Check the vehcile windows on both sides to determine the intensity of staining and how it occurred.*

A bead of fabric runs along each side of the window channels in most vehicles. The closer a window is to the seat of the fire, the less of this fabric will survive the fire. If only the fabric on the inside is burned away, the window was closed during an interior fire. If only the exterior fabric is burned away, the window was closed during an exterior fire. If the fabric on both sides of the window channel is burned away, either the window was open or both the inside and outside of the door were involved.

The positions of the windows are important: first, because windows may be opened to accelerate an interior fire; second, because there are norms as to window position, given the time of year, the weather conditions, and the location of the vehicle; and third, because the positions of the fire vehicle's windows will often substantiate or contradict the owner's statements.

Ignition Switch. The ignition switch should be checked for a key or the remains of a key. If the ignition key was left in the switch, an intense fire might melt off the protruding part of the key, which would then fall to the floor (perhaps along with other keys on a key ring). It might still resemble a key, or it might end up as a lump of melted-and-then-cooled metal. The part of the key that remained in the ignition switch should retain its shape.

Again, the presence or absence of the ignition key is important in substantiating the owner's statements—especially when the vehicle is reported as stolen.

Glove Compartment. The glove compartment lock should be checked for signs that it was forced open. Such evidence usually indicates theft, although the owner of the vehicle might force the lock to avert suspicion. The glove compartment itself should contain the usual items—repair bills and receipts, road maps, odd papers, the vehicle's registration, matches, and so on. These items may be useful in identifying the owner of the vehicle. A lack of paper items might indicate that they were used to start the fire.

Engine Compartment

The engine compartment contains very little flammable material. The insulation on electrical wiring will burn, as will deposits of grease on the engine, the small amount of gasoline in and around the carburetor, and perhaps the fan belt. But little else in the engine compartment will burn. For this reason, a fire in the engine compartment will usually burn itself out if the hood is closed; only rarely will such a fire reach the intensity needed to burn through the firewall and into the passenger compartment. However, if the hood is open, and the wind is blowing the fire against the windshield, then the windshield can break or melt. This will allow the fire to extend into the passenger compartment (Figure 10-10).

On the other hand, a fire in the passenger compartment, with plenty of fuel, could become intense enough to burn through the firewall and into

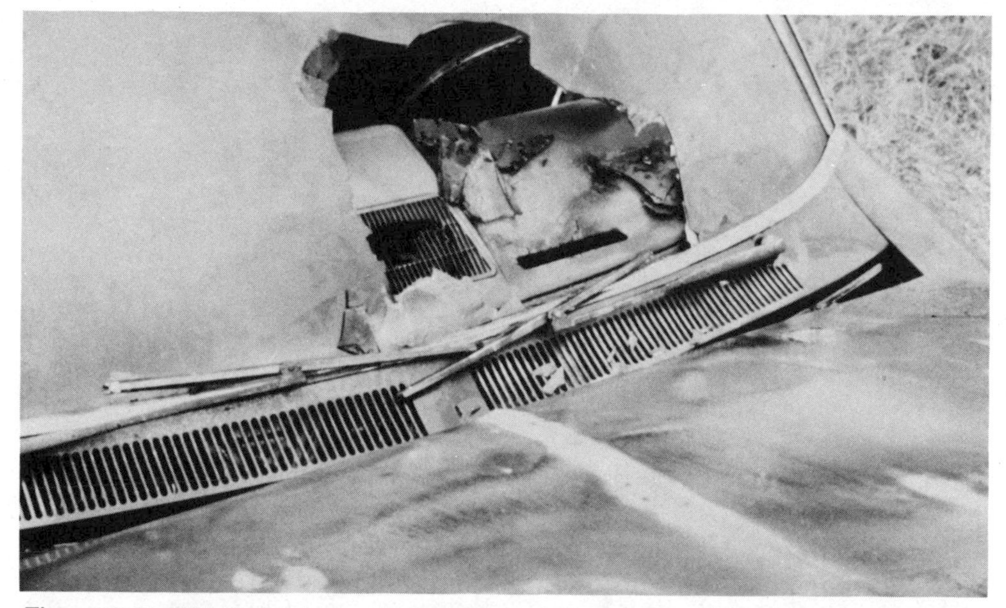

Figure 10-10. *With the hood open, the fire will extend into the passenger compartment (wind blowing against car).*

the engine compartment. If both these spaces were involved, the investigator must determine whether the fire spread from one to the other, or whether there were two separate points of origin.

Burn Patterns. The way in which the hood is damaged will indicate whether it was open or closed during an engine compartment fire. If the hood was open, the front of the hood will exhibit greater damage than the rear (Figure 10-11). If the hood was closed, the greatest hood damage will appear directly above the hottest part of the fire; the metal will probably be distorted in that area (Figure 10-12).

Figure 10-11. *When the wind is not a factor, an open hood front will have greater damage from flames and heat than the rear.*

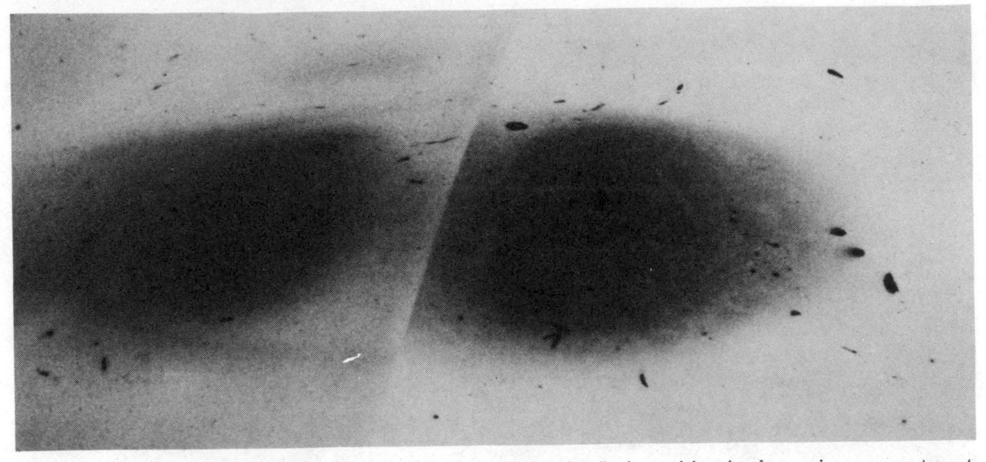

Figure 10-12. A closed hood shows heat damage above the fire's position in the engine compartment.

The carburetor is the most likely point of origin for a fire that originated above the engine. In most instances, the fire will not destroy the carburetor itself. Instead, burning will occur above and away from the carburetor; the burn patterns will show that the flames impinged heavily on the underside of the hood.

A fire that originates at or below the level of the engine will leave burn patterns on radiator hoses, fan belts, or motor mounts. It may melt solder on or around the radiator. The fire may also "clean" grease from the engine and other parts in the engine compartment. If these burn patterns indicate that the fire burned upward around the engine from below, it may well have been started with an accelerant on the ground.

Electrical Components. The electrical wiring should be examined carefully to determine whether the fire was electrical in nature (see Chapter 4). If the voltage regulator and distributor points are fused, the heat source may have been electricity. If a massive short occurred, the battery will be dead.

Engine. The head bolts on the engine should be examined. If they are loose, or if they show obvious tool marks, the engine may have been taken apart recently because of some operating problem. An engine in need of expensive repairs sometimes motivates vehicle arson.

The dipstick should be checked to determine the engine oil level. The absence of oil could indicate that the oil was drained before an attempt to repair the engine, but never replaced. Beads of water on the dipstick indicate that there is water in the oil pan; this in turn means that the engine block is cracked—again, sometimes a reason for abandoning or burning a vehicle. The engine block should then visually be checked for a crack.

Fuel Lines. Scratches or tool marks on fuel-line connections indicate that these lines have been tampered with. Disconnected fuel lines are obviously abnormal. If deposits of soot are found around the connectors and inside the lines, they were disconnected before the fire—probably to spill gasoline onto the engine.

Missing Parts. Generators, alternators, carburetors, starter motors, and batteries are expensive parts; they are easily sold, even in used condition. When such parts are missing from the fire vehicle, it is almost certain that the fire was deliberately set. In some cases the parts are stripped from the vehicle by thieves who then set the fire to try to cover the theft. In other cases, the owner salvages the parts before setting fire to his vehicle.

Automobile Trunk Space

The trunk of an automobile may contain a variety of flammable materials. However, the wiring that passes through the trunk to the rear lights is usually the only available heat source. For this reason, a fire that seems to have originated in the trunk should be carefully scrutinized; the investigator should make certain he has discovered the actual source of ignition.

Fire may spread from the passenger compartment to the trunk. However, when fire is found in the trunk and in the engine compartment but not in the passenger compartment, the obvious conclusion is that separate fires were set. This situation should be treated as a case of arson (Figure 10-13).

The trunk should be examined even when it was not involved with fire. The lock should be checked for signs of forcing (thieves sometimes punch the lock out completely). The contents should be checked: The trunk should contain at least a spare tire and a jack; most trunks contain a number of other items as well. An empty or partially empty trunk may have the same

Figure 10-13. *No fire in the passenger compartment, but in the engine compartment and trunk, could mean arson.*

meaning as missing engine parts—arson to cover theft or to get rid of an unwanted vehicle.

CASE HISTORY: Auto Fire

Firefighters responded to an auto fire at 7:01 a.m. The car was parked in a driveway alongside a single-family dwelling. The passenger compartment was well involved with orange-red flames. Seconds after firefighters arrived, the car windows seemed to fail all at once and fall out; at this point, gray-black smoke issued from the car.

After the fire was extinguished, the senior fireground officer spoke to the owner of the automobile. He said he had awakened at about 5:30 a.m. and had looked out in the general direction of the driveway; the car was not burning at that time. He returned to bed and was awakened at about 7 a.m. by a passer-by who told him the car was on fire. He also said he had not used the car since noon on the preceding day. The officer called for a fire investigator.

The investigator considered the smoke and flame colors as indicating the burning of normal combustibles (for a vehicle). The failure of the windows indicated a long burn time; this was supported by the thick coating of smoke on the window fragments. Exterior examination indicated that the outside of the car was involved only minimally. The car roof was distorted very slightly, above the driver's seat. The engine compartment was not involved, and was normal in all respects. There was minor burning in the trunk, where fire had spread from the rear seat of the passenger compartment; the contents of the trunk were in order.

The passenger compartment was burned heavily but very evenly. The handles and window levers were damaged to about the same extent on both front doors. The springs on the front seat were partially collapsed, more so on the driver's side. The rear seat springs retained their elasticity. The fire seemed to have originated on or near the front seat, but no heat source or single point of origin was visible. The investigator felt certain that the fire had been set deliberately.

Further questioning of the owner revealed that he had recently reported the newsboy who delivered his morning paper; the boy had stolen money that was meant for a second newsboy. When the boy was questioned, he admitted that he had almost lost his job because the theft was reported, and had set the fire to get even. He had entered the unlocked car when he delivered the paper at about 6 a.m. and had found a box of facial tissues on the front seat. He then piled the tissues along the seat, ignited them with a match, and left.

This case of arson was uncovered because the vehicle fire was investigated as thoroughly as a structure fire. The investigator conducted an exterior examination and an interior examination and questioned available witnesses. When his investigation failed to yield a point of origin and heat source, he investigated further until he was able to categorize the fire. He did not rush to the conclusion that a vehicle fire with no signs of accelerant use must have been caused by careless smoking.

11

DOCUMENTS AND REPORTS

Occasionally, a fireground investigation is delayed until some time after firefighting operations have been completed. It may then be necessary for the investigator to obtain permission to enter the fireground. The circumstances in which permission is required, and the documents needed to obtain this permission, are discussed in the first part of this chapter. The remainder of the chapter is devoted mainly to the Fire Investigation Report, which is completed at the end of the investigation and which serves to tie the entire investigation together.

DOCUMENTS RELATED TO DELAYED ENTRY*

The Fourth Amendment to the Constitution of the United States protects its citizens and their property against "unreasonable searches and seizures." It requires that a search warrant be issued only "upon probable cause, supported by oath or affirmation, and particularly describing the place to be searched, and the persons or things to be seized." It is this amendment which, for example, prohibits police from arbitrarily entering and searching a person's home or business, confiscating his property, or arresting him without cause.

---

*This section draws in part on the article "Fire Scene Examination and the Fourth Amendment" by William T. Kennedy, in *The Fire and Arson Investigator,* vol. 31, no. 1, 1980, and *Arson Detection and Investigation for Fire Officers,* developed by the Maryland Fire and Rescue Institute.

In 1978, in its decision in the case of *Michigan versus Tyler,* the United States Supreme Court clarified the provisions of the Fourth Amendment as they apply to fireground investigation.

Michigan v. *Tyler*

Fire companies arrived at a fire involving a furniture store in Oakland County, Michigan, around midnight, and proceeded with firefighting operations. At about 2 a.m., during overhaul, the fire chief arrived at the scene. He was notified that two containers of flammable liquid had been found. It was difficult to photograph the scene because of the smoke, steam, and darkness. At about 4 a.m., all firefighters left the scene; the chief took the two containers and turned them over to police.

At about 8 a.m., the chief returned to the scene (which was unattended) with a fire investigator; they examined the scene briefly and then left. At 9 a.m., the chief returned to the scene, this time with a detective. They examined the structure carefully, took photographs, and discovered suspicious burn patterns that had not been visible in the darkness. They took samples from a carpet and a stairway, and searched through the debris for additional evidence. They then left the scene with this evidence.

The next day, and during the next month, fire and police officials visited the scene, taking additional photographs and physical evidence of arson.

The store owners—two partners—were eventually charged with arson and brought to trial. The defense claimed that the evidence found at the scene could not be used at the trial because it was obtained without a search warrant, in violation of the Fourth Amendment. The court disagreed, and the defendants were found guilty. They appealed to the Michigan Supreme Court, where the lower court ruling was reversed. The state of Michigan then appealed to the United States Supreme Court.

The United States Supreme Court agreed, in principle, with the ruling of the Michigan Supreme Court. The defendants were set free. The Supreme Court ruled that the defendants' rights under the Fourth Amendment had been violated when officials searched the fire scene without search warrants. In particular the Court noted that:

1. No search warrant is needed when fire department personnel enter a structure to fight a fire.
2. Once fire personnel are in a structure, they may remain there for a reasonable amount of time after extinguishment to determine the cause of the fire. They may leave temporarily and return within that reasonable amount of time.
3. For re-entry after a reasonable amount of time, a search warrant is required unless the occupant has consented to a search of the premises.

The United States Supreme Court decision implies that investigators should consider a "reasonable amount of time" to be no more than four hours. (The Court ruled that the entries made by the chief at 8 a.m. and 9 a.m. were within a reasonable time of extinguishment, but later entries were not.) If investigators need to enter the fire structure more than four hours after firefighting operations have been completed, they should attempt to obtain the occupant's consent; otherwise, any evidence they find may not be admitted in court. If the occupant is unavailable or refuses to sign a consent form, the investigator must obtain a search warrant. Either of two types of warrants may be required, depending on the reason for the search.

Consent to Search Form

The decision in *Michigan versus Tyler* indicates that the occupant's presence during a delayed investigation does not imply that he has consented to the search. The occupant must consent to the delayed investigation in writing. Further, he must be made aware of (1) his constitutional right to refuse entry to investigators and (2) the fact that any evidence found during the search may be seized and used against him in court.

The model consent form shown on page 255 complies with these requirements. The occupant should be asked to sign such a form whenever fire department personnel expect to re-enter the fire structure to begin or continue an investigation. When it is known that the investigation will be delayed, the fireground officer should try to obtain a signed consent form before fire companies leave the scene to return to quarters.

Administrative Search Warrant

When the occupant of the fire structure is unavailable or refuses to sign a consent form, investigators must obtain a search warrant from a judge or magistrate. Without the warrant, they cannot legally enter the structure, if such entry is to be made more than four hours (a reasonable amount of time) after extinguishment. If entry is required only to perform a fireground investigation—and not because investigators believe a crime was committed—they may obtain an administrative (or special) search warrant.

The requirements for obtaining an administrative search warrant are less stringent than those for a criminal search warrant. In many jurisdictions, the investigator must show only that a fire has occurred and that the fire investigation was not completed. In other jurisdictions, the investigator must also convince a judge that the search will be reasonable; give the specific time at which the search will be conducted; disclose the number of entries that have been made previously; or provide other relevant information.

A typical administrative search warrant (here called a special inspection warrant) is shown on page 256, in addition to the affidavit on page 257 that the investigator must complete and swear to, in order to obtain the warrant.

The administrative search warrant limits the investigation to the determination of the point of origin, heat source, reason, and category of the fire. However, if evidence of arson is found during this investigation, it may be collected and confiscated—just as if the investigation had not been delayed.

Because an administrative search warrant is easier to obtain than a criminal search warrant, it does not fully comply with the provisions of the Fourth Amendment. Therefore, it may not be used to search for evidence when investigators already have reason to believe a crime was committed; if it is so used, the evidence that is found may not be allowed in court. When investigators believe a crime was committed, they must obtain a criminal search warrant.

Criminal Search Warrant

A criminal search warrant is required when fire investigators believe arson was committed and need to enter the fire structure to search for additional evidence. To get the warrant, the investigators must show, to a judge, that there is "probable cause" to believe both that a crime was committed and that evidence of the crime will be found in the fire structure. Normally, criminal search warrants are obtained by arson investigators; the involvement of the fireground investigator ends when he has determined the category of the fire.

Delayed Investigation

Delayed investigation of the fireground may be more difficult than investigation beginning during or immediately after fire suppression. However, it is far from impossible. Insurance investigators, for example, normally begin their investigations several days after extinguishment, and they have a high rate of success.

Once fire department personnel leave the fire structure, an arsonist may be able to re-enter and remove or disguise evidence. But it is impossible to alter burn patterns and other evidence of the origin and spread of the fire, unless the structure is destroyed completely. In addition, fire department reports contain much information concerning the fire—information that can be compared with what is found during the delayed examination.

There is also the inconvenience of obtaining a signed consent form from the occupant, or an administrative search warrant. This procedure, however, soon becomes routine. Moreover, if the time period between the end of firefighting operations and the beginning of the fireground investigation is expected to be only a few hours, the fire department may be able to maintain custody of the structure. This is done simply by leaving department personnel at the scene. The fireground investigation must still begin within a "reasonable amount of time" after extinguishment; however, now the fire department (and not the occupant) has control of the building during the time between extinguishment and investigation.

The *Michigan versus Tyler* decision did not deal with the custody of the fire building during the "reasonable amount of time." For this reason, fireground investigators and company officers should be aware of state and local laws concerning the custody of structures involved in fires. If the fire department has the right to maintain custody and does so, then:

1. The investigator need not obtain either a consent form or a search warrant.
2. Unauthorized persons (including occupants) may be prohibited from entering the fireground until the investigators arrive, perform their duties, and leave.
3. The investigators can assume that nothing on the fireground was disturbed in the time since firefighting operations were completed.

FIRE DEPARTMENT REPORTS

The reports filed by the fire department dispatcher, the chief officer, and company officers can provide information concerning the

- Time of the alarm
- Points of entry
- Conditions and positions of doors and windows prior to entry
- Colors, locations, amounts, and patterns of smoke and flames
- Weather conditions during the fire
- Specific firefighting actions that were taken
- Unusual occurrences during firefighting operations.

In addition, the reports list the responding companies and the names of responding officers and firefighters.

The investigator who arrives during firefighting operations can question firefighters and officers at the scene. He may, however, want to read these reports before completing his investigation. When the investigation is delayed, he should read through the fire department reports before going to the fireground. In either case, he should consult with the officers who wrote the reports if anything is unclear or if he believes the officers or men can provide additional information.

Samples of fire department reports are included in the case history at the end of this chapter.

FIRE INVESTIGATION REPORT

Once the point of origin, heat source, reason, and category of the fire have been determined, the investigation is nearly over. If the category was determined to be arson, the scene would have been turned over to the proper authority according to the procedures detailed by state and local ordinances. All evidence would have been made available to that authority. Whatever the category of the fire, only the Fire Investigation Report remains to be written.

Most fire departments supply report forms to be filled out by the investigator. A model form is shown on page 258. The form should be simple, and it need not require excessive detail or effort; the details of the investigation should be contained in the investigator's notes and tape recordings, diagrams, and photos.

The information that is placed in the report should be confined to facts, and it should be accurate. Supplementary sheets may be added to forms such as the one shown on page 259 if they are needed to complete the "Supporting Data and Remarks" section, or to describe physical evidence that was collected during the investigation.

Copies of the report should be sent to fire department personnel as prescribed by department policy. In many departments the fire chief's office—and not the individual investigator—disseminates Fire Investigation Reports to interested outside agencies.

The original report should be retained on file—preferably in or near the locked storage room that is reserved for evidence collected by investigators. If the statements of occupants and witnesses are transcribed from tape, the transcriptions should be filed with the original Fire Investigation Report. (The tapes themselves would be stored with other evidence pertaining to the investigation.) It is also a good idea to file, with the original report

- Copies of fire department reports pertaining to the fire
- Copies or originals of any laboratory reports or experts' reports that were requested as part of the investigation
- Relevant warrants or consent to search forms and copies of affidavits sworn by investigators
- The diagrams and photos produced during the investigation
- The originals or copies of any other documents related to the investigation and the investigator's determinations.

A sample checksheet that may be used to keep track of the documents and evidence contained in the investigation "package" is shown on page 260.

CASE HISTORY: Firefighting and Investigative Reports

At 2:30 a.m. in late November, the Fire Department was dispatched to a tavern in a one-story building, with living quarters in the rear of the structure. Attack crews made the following observations on arrival and during fire suppression:

Observation	Indication
• Fire showing in the window of the barroom with orange-yellow flames with red intermingled. Smoke grey with some black smoke.	• Indicates ordinary combustibles and hydrocarbon materials involved
• Two occupants outside wrapped in blankets	• Occupants presumed to be retired before fire
• Side door of barroom standing open	• Possible illegal entry
• Some difficulty in gaining extinguishment	• Possibly a high combustible substance involved.

Following extinguishment, the fireground officer began an initial investigation. It was apparent that the fire originated behind the bar at floor level. As he examined the point of origin, he found a melted waste container containing remains of cigarette butts and paper bar napkins. He did, however, detect a strong odor of kerosene. At this time he called for an arson investigator. He immediately ordered the scene to be completely protected as a possible crime scene.

The fire/arson squad arrived and began gathering evidence. It was called to their attention the fact that attack crew of engine 6 found the side door ajar. Close examination of the door confirmed suspicion that there was illegal entry. The door had been pryed with a flat tool and popped open. It was also confirmed that kerosene had been poured under the bar. Samples were taken for analysis. Further search of the point of origin determined that there was no other possible heat source that could have caused ignition other than the cigarette butts or direct contact by open flame. However, it was determined that the kerosene was placed at the point of origin and ignited to make it appear that the cigarette butts ignited an accidental fire.

Search of the surrounding area proved productive with the turning up of a large glass fruit jar containing odor of kerosene. It apparently was dropped in the grass by the person involved while leaving the scene. This jar will be checked for fingerprints. It was impossible to provide any footprints outside. All evidence was photographed before being moved. The occupants were interviewed. The man was the owner of the tavern and lived in the rear living quarters. The woman was a hired bartender. During questioning as to any problems with people, it came out that the woman bartender had very recently broken off with a boyfriend. She stated he was very belligerent about it, however, he had not threatened anyone.

The following reports are involved in the fire situation:

1. Chief Officer's Report .... see page 261
2. Engine-6-Officer's Report .... see page 262

3. Engine-7-Officer's Report .... see page 263
4. Track-9-Officer's Report .... see page 264
5. Dispatchers Report .... see page 265
6. Fire Investigation Report .... see page 266
7. Photography List .... see page 267
8. Evidence Record .... see page 268

At times, other reports may be required depending on the circumstances of each individual case.

Figure 11-1, a diagram of a fire scene, should also be included with the fire situation reports.

Every fire investigation ends in the writing and compilation of reports to preserve facts and identify evidence. These documents may never be used in litigation, but, on the other hand, they may become the backbone of evidence to bring the career of an arsonist to its end.

CONSENT FORM FOR FIRE SCENE EXAMINATION

I, ______________________________ have been requested to consent to
(name)

an examination of my property located at ______________________________

____

____
(full description and address of property)

This examination is being conducted as part of an investigation of the fire which occurred to this property on ______________________________
(date of fire)

I am the lawful occupant of this property. I have been advised of my constitutional rights to refuse such consent and to require that a search warrant be obtained prior to any examination. I have further been advised that if I do consent to an examination, any evidence found as a result of such examination can be seized and used against me in any court of law, and that I may withdraw my consent at any time prior to the conclusion of the examination.

After having been advised of my constitutional rights as stated above, I hereby voluntarily waive those rights and consent to an examination and authorize:

____
(names of town or city Fire Investigators or their representatives)

to conduct a complete examination of the described property. Further, permission is granted to remove from this property any material deemed pertinent to the investigation of this fire.

____
(signed) (date)

Witnesses:

____

____

____

SPECIAL INSPECTION WARRANT

STATE OF WISCONSIN
COUNTY OF WINNEBAGO } ss

In the ____________________ Court of the County of Winnebago

The State of Wisconsin, to the sheriff or any constable or any peace officer of said county:

Whereas ____________________________ has this day complained in writing to said court upon oath that on the ____ day of ______________ 19____ in said county, in and upon certain premises more particularly described as follows: ____

____

there now exists a necessity to determine the cause, origin, and circumstances of a fire in compliance with Section 165.55(1) and prayed that a special inspection warrant be issued to search said premises.

Now therefore in the name of the State of Wisconsin, you are commanded forthwith to search the said premises for said purposes.

Dated this _____ day of _______________ 19_____

____

Judge of the _______________ Court

Endorsement of Warrant

Received by me ___________________ 19_____ at _____ o'clock _____m.

____

Inspector/Investigator

AFFIDAVIT

STATE OF WISCONSIN }
COUNTY OF WINNEBAGO } ss

In the ____________________ court of one county of ____________________,

an investigator for the ____________________ Fire Department, being duly

sworn, says that on the _____ day of ________________ 19_____ in said

county, in and upon certain premises more particularly described as follows:

____

____

there now exists a necessity to determine the cause, origin, and circumstances of a fire which occurred on said premises in compliance with Section 165.55(1); Wisconsin Statutes. The facts tending to establish the grounds for issuing a special inspection warrant are as follows:

1. The premises were damaged by fire on or about the _______ day of

   ____________________ 19_____.

2. Under the provisions of Section 165.55(1), ____________________ Fire Chief and his designated investigators are authorized to investigate fires which may be of incendiary origin.

3. An investigation of the fire pursuant to Section 165.55(1) has not been completed.

4. On the _____ day of ________________ 19_____ at ________________

   _____ m. ______________________, the person presently having control of the premises did refuse consent to enter the premises for the purpose of conducting said investigation. Wherefore, the said investigator prays that a special inspection warrant be issued to search such premises for said purpose.

Signed ____________________

Subscribed and sworn to before me this _____ day of ________________

19_____. ______________________________

Judge of the ________________ Court

PRINCE GEORGE'S COUNTY, MARYLAND
FIRE DEPARTMENT
BUREAU OF FIRE PREVENTION
INVESTIGATION REPORT

1	DATE/TIME OCCURRED	WEATHER	F.R. / CCN
2	LOCATION OF INCIDENT		
3	T/A	ADDRESS	
4	TYPE OF INCIDENT ☐ STRUCTURE FIRE ☐ FALSE ALARM ☐ POINT OF ORIGIN ______ ☐ OUTSIDE FIRE ☐ MOLOTOV ☐ VEHICLE FIRE ☐ EXPLOSIVE DEVICE ☐ HEAT SOURCE ______		
5	M/O		
6	ESTIMATED $ LOSS	STRUCTURE	CONTENTS / OTHER
7	COMPLAINANT'S NAME	RACE-SEX / AGE-D.O.B.	
	ADDRESS	PHONE	
8	REPORTING PERSON	RACE-SEX / AGE-D.O.B.	
	ADDRESS	PHONE	
9	INDICATE: ☐ OWNER ☐ OCCUPANT ☐ VICTIM ☐ SUSPECT ☐ OTHER (specify)		
	NAME	RACE-SEX / AGE-D.O.B.	
	ADDRESS	SS. NO. / PHONE	
	OCCUPATION ADDRESS	PHONE	
10	INDICATE: ☐ OWNER ☐ OCCUPANT ☐ VICTIM ☐ SUSPECT ☐ OTHER (specify)		
	NAME	RACE-SEX / AGE-D.O.B.	
	ADDRESS	S.S. NO. / PHONE	
	OCCUPATION ADDRESS	PHONE	
11	INDICATE: ☐ OWNER ☐ OCCUPANT ☐ VICTIM ☐ SUSPECT ☐ OTHER (specify)		
	NAME	RACE-SEX / AGE-D.O.B.	
	ADDRESS	S.S. NO. / PHONE	
	OCCUPATION ADDRESS	PHONE	
12	ACTION TAKEN ☐ THIS REPORT ☐ M.I.R.S. ☐ EVIDENCE ☐ J-1 ☐ OTHER (specify) ☐ SUPP. REPORT ☐ PHOTOS ☐ WARRANT ☐ J-2		
13	CASE STATUS	IF CLOSED SPECIFY HOW	
14	INVESTIGATOR I.D.	BUREAU O.I.C.	

DETAILS ON REVERSE

A report form containing at least the following information:

General:
Alarm (or incident) number
Date of alarm
Time of alarm
Date of investigation
Time of investigation
Investigator(s)
Photos taken by
Photos processed by
Diagrams drawn by

Structure:
Address
Type and occupancy
Type of construction
Structure height, length, width
Owner
Occupant

Vehicle:
Make and model
Year
Color
Registration number
VIN
Owner

Determination:
Point of origin
Heat source
Reason
Category

Supportive Data and Remarks

____

____

____

____

CHECKLIST/REPORTS AND DOCUMENTS

Reports	*Attached*	
	Yes	No
Fire Department	________	________
Police Department	________	________
Coroner	________	________
Dispatcher	________	________
Laboratory	________	________
Consent to search	________	________
Warrants	________	________
Technical experts	________	________
Other	________	________

Statements: Taken by ____________________

Given by ____________________

Attached Yes _____ No _____

Diagrams: Drawn by __________ Number _____

Attached Yes _____ No _____

Evidence: Secured Yes _____ No _____

Disposition: ____________________

Types: ____________________

Date: __________

CHIEF OFFICER'S REPORT

ALARM NUMBER 1104 DATE 11-23 19 80

LOCATION OF ALARM 432 Cresent Road

DELAYED: ___ YES X NO REASON: ___ TIME 2:30 A.M.

ALARM DISCOVERED BY: George Able ADDRESS 432 Cresent Road

ALARM REPORTED BY: George Able ADDRESS 432 Cresent Road

RESPONDED FROM: Station 1 ON DUTY MEN 10

COMPANIES RESPONDING: Engine 6 & 7, Truck 9 OFF DUTY MEN 0

TYPE OF BLDG: Tavern & Residence HT 1 story CONSTRUCTION Wood Frame

OWNER: George Able ADDRESS 432 Cresent Road

OCCUP: George Able ADDRESS 432 Cresent Road

OCCUP: Amy Stang ADDRESS 432 Cresent Road

INSURANCE CARRIER: Olympic Mutual

DAMAGE:

	FIRE		SMOKE	
STRUCTURE:	MAJOR ___	MINOR ___	MAJOR ___	MINOR ___
CONTENTS:	MAJOR ___	MINOR ___	MAJOR ___	MINOR ___

WEATHER CONDITIONS:

TEMP ___ WIND ___ VELOCITY: ___ LT ___ MED ___ STR

WEATHER AND ROAD CONDITIONS Dry – cold

OPERATIONS: While enroute, Engine 6 notified me that there was a working fire in the barroom. I called Engine 7 and Truck 9 to prepare to wear BAs for entry and rescue. As I arrived, Engine 6 was pulling an attack line. I told Truck 9 to ventilate the roof above the fire and to perform search and rescue in the living quarters. Engine 6 did extinguish the fire with some difficulty which now was because of kerosene being present. The occupants were already out of the building. Following extinguishment, I began an initial fire investigation. The point of origin was obvious directly behind the bar at floor level, between the glass washer sink and the beer coolers. There was a melted plastic container with remains of cigarette butts and paper. There was a strong odor of kerosene. Immediately I ordered the area protected and for the fire/arson squad to make an arson investigation. Overhaul was performed carefully to avoid destroying any possible evidence. The investigation team arrived. I sent Engine 6 and Truck 9 back to quarters and kept Engine 7 on the scene to assist the investigation team. I filled them in on the information I had and returned to quarters.

Rexford Clune — Chief Officer

COMPANY OFFICER'S REPORT
GENERAL ALARM

ALARM NUMBER 1104 COMPANY Eng. 6 DATE 11-23- 19 80

LOCATION OF ALARM 432 Cresent Road

RESPONDED FROM Station 3

DELAYED: YES ___ NO X REASON ___ TIME 2:30 A.M.

EQUIPMENT USED:

EXTINGUISHERS: CO_2 ___ ABC ___ H_2 ___ HALON ___

PUMP: YES ___ NO X TIME: ___ HRS 4 MIN

HOSE:

SIZE HOSE	FOOTAGE WET	FOOTAGE DRY	FROM	TO	PUMP PRESSURE	ESTIMATED GALLONS
1"						
1½"	200		Pump	Fire	150	400
2½"						
3"						

OTHER:

QUANTITY	EQUIPMENT	DAMAGED	LOST
3	Breathing apparatus		

OPERATIONS: On arrival, we could see smoke and fire in the bar area. Flames were yellow-orange with some red showing, smoke was grey with black mixed in. We could see the fire appeared to be in the SW corner of the barroom. I radioed the Chief Officer and informed him of the situation. We pulled a 1½" preconnect line and made entry in the side door on the NE corner. NOTE: The side door was ajar about 3". We pulled the line and were able to attack the fire. It seemed the fire spread some and was difficult to extinguish. When the fire was out we removed our line and reported the door being open and the difficulty with extinguishment. We then returned to quarters.

RESPONDING PERSONNEL: Lt. John Dopus

Driver Jim Loper

Firefighter Lange Hope

OFFICER IN CHARGE: Lt. J. Dopus

COMPANY OFFICER'S REPORT
GENERAL ALARM

ALARM NUMBER 1104 COMPANY Eng. 7 DATE 11-23- 19 80

LOCATION OF ALARM 432 Cresent Road

RESPONDED FROM Station 2

DELAYED: YES ___ NO X REASON ___ TIME 2:30 A.M.

EQUIPMENT USED:

EXTINGUISHERS: CO_2 ___ ABC ___ H_2 ___ HALON ___

PUMP: YES ___ NO X TIME: ___ HRS ___ MIN

HOSE:

SIZE HOSE	FOOTAGE WET	FOOTAGE DRY	FROM	TO	PUMP PRESSURE	ESTIMATED GALLONS
1"						
1½"						
2½"	300		hydrant	Eng. 6		
3"						

OTHER:

QUANTITY	EQUIPMENT	DAMAGED	LOST
2	Breathing apparatus		
1	Ax		
1	Short pike pole		

OPERATIONS: When we arrived on the scene, Eng. 6 was pulling an attack line to the building. We dropped a 2'½" line from the hydrant and connected to Engine 6. Myself and firefighter Stang donned BAs and followed the line to help Engine 6. My driver helped Truck 9. As we entered the fire was extinguished. We assisted Truck 9 with overhaul. We picked up our equipment and stayed on the scene to help the investigators.

RESPONDING PERSONNEL: Captain Newcomb Last

Driver Ron Birch

Firefighter Leland Stang

OFFICER IN CHARGE: Capt. N. Last

COMPANY OFFICER'S REPORT
GENERAL ALARM

ALARM NUMBER 1104 COMPANY Trk. 9 DATE 11-23- 19 80

LOCATION OF ALARM 432 Cresent Road

RESPONDED FROM Station 2

DELAYED: YES ___ NO X REASON ________ TIME 2:30 A.M.

EQUIPMENT USED:

EXTINGUISHERS: CO_2 ___ ABC ___ H_2 ___ HALON ___

PUMP: YES ___ NO ___ TIME: ___ HRS ___ MIN N/A

HOSE:

SIZE HOSE	FOOTAGE WET	FOOTAGE DRY	FROM	TO	PUMP PRESSURE	ESTIMATED GALLONS
1"						
1½"						
2½"			N/A			
3"						

OTHER:

QUANTITY	EQUIPMENT	DAMAGED	LOST
2	24' ladders—2 AX-2 Pike poles		
3	500 Watt lights		
4	Breathing apparatus		
6	12 x 12 tarps		
1	6 x 6 plastic lathe for roof cover		
1	K-12 saw		

OPERATIONS: Upon arrival, Engine 6 was pulling line for attack. The Chief Officer told us to ventilate the building vertically. Two of us proceeded to the roof and cut a 4 x 4' hole above the fire and relieved the smoke and gases. Two of us went to search the living quarters. Two occupants, a man and a woman wrapped in blankets, informed them that everyone was out. They proceeded to open windows in the living quarters to vent products of combustion. Engine 6 made entry. We set up lighting and generator. Driver Ron Birch from Engine 7 worked with my crew. We donned BAs and spread tarps in the barroom. Following extinguishment, we were ordered to overhaul the scene. Engine 7 assisted us. We patched the roof temporarily, picked up, and returned to quarters.

RESPONDING PERSONNEL: Lt. Emil Harth

Driver Jake Hanna — Fill-in from Eng. 6—Driver Ron Birch

Firefighter Allan Huckster — Firefighter James Forth

OFFICER IN CHARGE: Lt. E. Harth

DISPATCHER'S REPORT

ALARM NUMBER 1104 DATE November 23 19 80

TEMP 29 deg. F WIND DIRECTION N WIND VELOCITY 11

ALARM INFORMATION

TIME—1ST ALARM ________ CO'S DISPATCHED ________

TIME—2ND ALARM ________ CO'S DISPATCHED ________

TIME—3RD ALARM ________ CO'S DISPATCHED ________

TIME UNDER CONTROL 2:42 A.M.

TYPE OF ALARM RECEIVED Telephone

METHOD OF DISPATCH Radio

LOCATION OF ALARM 432 Cresent Road

REASON Arson—Man was jealous when his girlfriend jilted him & moved in with George Able

OWNER George Able ADDRESS 432 Cresent Road

OCCUPANT George Able and Amy Stang ADDRESS 432 Cresent Road

FIRE DAMAGE Major SMOKE DAMAGE Major

EQUIPMENT FAILURE

N/A

CHIEF ON DUTY Rexford Clune/Assistant Chief

ADDITIONAL REMARKS: Chief Clune requested the fire/arson squad at 2:48 A.M. The fire/arson squad arrived on the scene at 3:05 A.M.

ALARM OPERATOR Bary Oestrick

FIRE/INVESTIGATION REPORT

(1) Alarm No. 1104 (2) Date 11-23 19 80

(2) Temperature 29 degrees (4) Wind Direction N-11

(5) Time of Alarm 2:30 AM (6) Time of Investigation 3:05 AM

(7) Location of Alarm 432 Cresent Road

(8) Alarm discovered by George Able

(9) Alarm reported by George Able

(10) Owner George Able (11) Address 432 Cresent Road

(12) Occupant George Able & Amy Stang (13) Address 432 Cresent Road

(14) Type of building Tavern and Residence (15) Height 1 Story

(16) Construction Wood Frame

(17) Photos taken by Jamie Rath (18) Date 11-23-80 (19) Time 3:05 AM

(20) Category Arson

(21) Reason Man was jealous when his girlfriend jilted him and moved in with George Able.

(22) Heat Source Direct contact

(23) Investigative Remarks We arrived at the fire scene and met Chief Officer Rexford Clune. He informed us of his findings. In checking the point of origin, we took samples of the floor that had a strong odor of kerosene. The floor showed a burn pattern of a flammable or combustible liquid. All possible heat sources were checked and eliminated with the exception of direct ignition by a match or similar device. The arsonist started the fire at this point to make it appear as an accidental fire ignited by the discarded cigarette butts and paper. The door on the north side of the building was pryed open. Search of exterior turned up a glass 2 quart jar that had residue of kerosene inside. This was also sent to the laboratory for possible fingerprints. An interview with George Able and Amy Stang revealed that her former boyfriend, William Fromm, was bitter and upset when she dropped him and moved in with George Able.

11/24/80 — Contacted William Fromm and interviewed him. He denied any connection, however, he could not prove his whereabouts at the time the fire occurred. When we confronted him with his fingerprints on the glass jar, he admitted setting the fire.

(24) Investigators: Alex Woodruf

Jamie Rath

FIRE INVESTIGATION—432 Cresent Road—11/23/80

PHOTO LIST

Photo 1
Exterior of west side of structure

Photo 2
Exterior of north side of structure

Photo 3
Exterior of east side of structure

Photo 4
Exterior of south side of structure

Photo 5
Wide angle of glass fruit jar north of building

Photo 6
Closeup of glass fruit jar north of building

Photo 7
Closeup of door on north side—shows pry marks

Photo 8
Picture of roof—ventilation hole

Photo 9
Closeup of window west side—barroom—medium staining. Oily film

Photo 10
Overall view of bar from front door SW corner shows damage of bar

Photo 11
The front area of the bar which was damaged the most. It is looking from the patrons side and indicates heavy charring on upright members and bar stools

Photo 12
Front of bar looking at exposed back of cooler. Note the discoloration and distortion of the metal

Photo 13
Flame pattern on bar stools

Photo 14
Bottom of bar on the patron side showing flame burn through from beneath bar. Note flame and heat damage to cooler

Photo 15
Overall view of heavy burned area showing distortion of cooler

Photo 16
Loo[illegible] patron side u[illegible] was taken ne[illegible]he
[illegible]

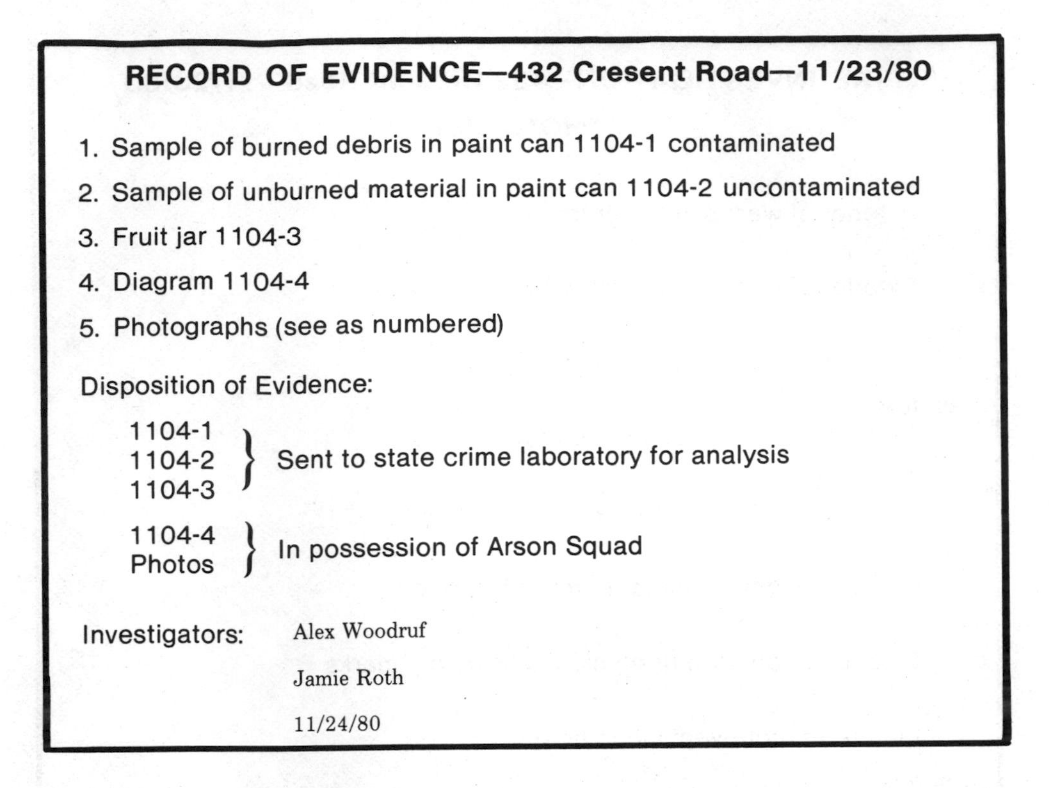

RECORD OF EVIDENCE—432 Cresent Road—11/23/80

1. Sample of burned debris in paint can 1104-1 contaminated
2. Sample of unburned material in paint can 1104-2 uncontaminated
3. Fruit jar 1104-3
4. Diagram 1104-4
5. Photographs (see as numbered)

Disposition of Evidence:

1104-1
1104-2
1104-3 } Sent to state crime laboratory for analysis

1104-4
Photos } In possession of Arson Squad

Investigators: Alex Woodruf
Jamie Roth
11/24/80

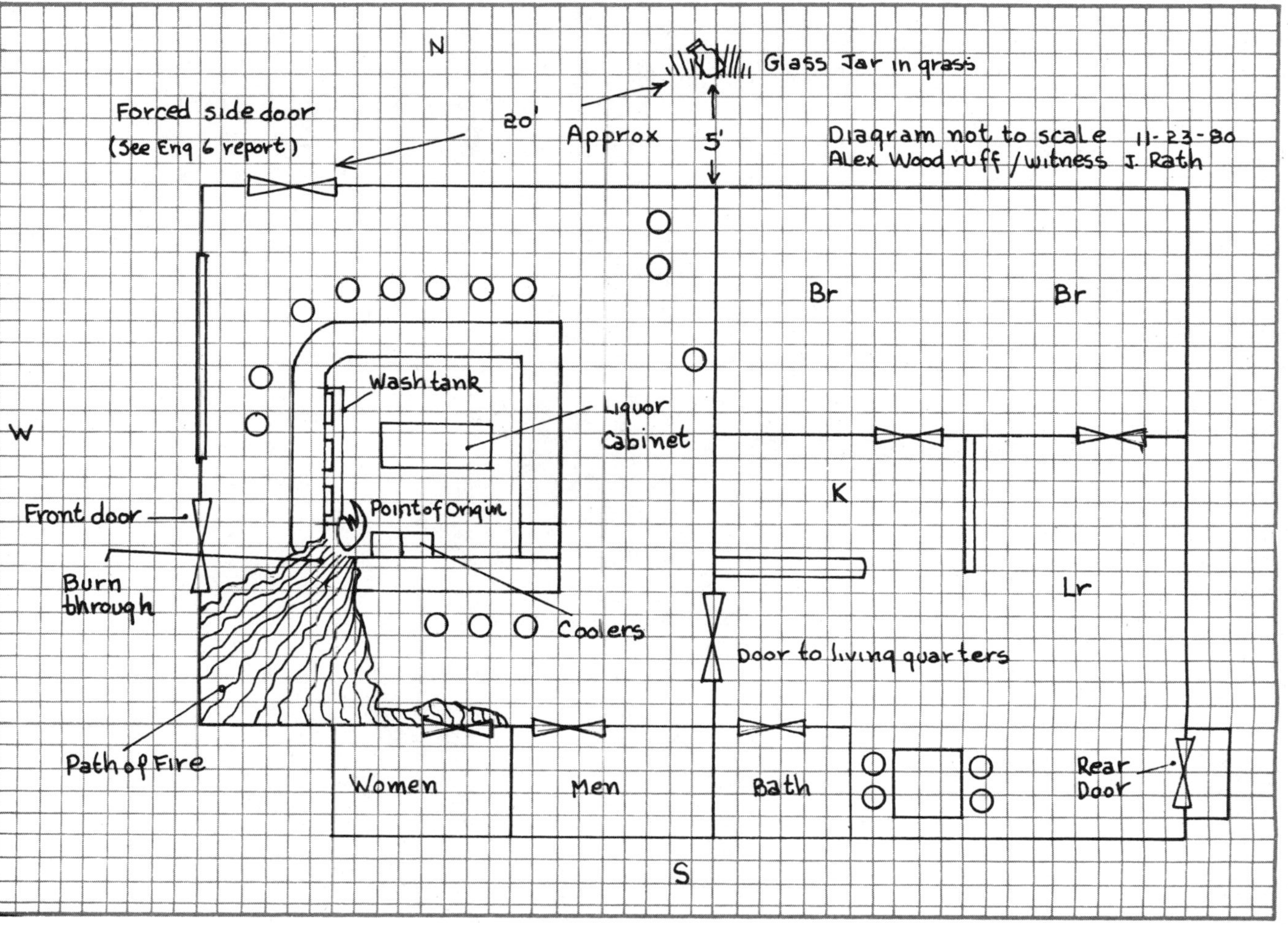

Figure 11-1. Diagram of a fire scene.

INDEX

C

I

J

K

L

M

N

O

P

W